ESTIMATING
FOR
HEATING AND
VENTILATING

Books of allied interest

HOT WATER SUPPLY—Design and Practice
  by J. J. Barton, C.Eng., M.Inst.F., F.I.H.V.E., Hon. I.D.H.E.

HANDBOOK OF HEATING, VENTILATING AND AIR CONDITIONING
  by John Porges, M.I.Mech.E., M.Inst.F. (Revised by F. Porges, LL.B., B.Sc. Eng., C.Eng., M.I.Mech.E., M.I.H.V.E.)

HEATING AND VENTILATING—Principles and Practice
  edited by J. J. Barton, C.Eng., M.Inst.F., F.I.H.V.E., Hon. I.D.H.E.

ELECTRIC FLOOR WARMING—With Notes on Ceiling Heating
  by J. J. Barton, C.Eng., M.Inst.F., F.I.H.V.E., Hon. I.D.H.E.

SMALL BORE HEATING AND HOT WATER SUPPLY FOR SMALL DWELLINGS—2nd Edition (Metric)
  by J. J. Barton, C.Eng., M.Inst.F., F.I.H.V.E., Hon. I.D.H.E.

ESTIMATING THE HEAT REQUIREMENTS FOR DOMESTIC BUILDINGS
  by J. J. Barton, C.Eng., M.Inst.F., F.I.H.V.E., Hon. I.D.H.E.

ESTIMATING

FOR

HEATING AND VENTILATING

J. J. BARTON

C.Eng., M.Inst.F., F.I.H.V.E., Hon. I.D.H.E.

(Consultant Heating and Ventilating Engineer)

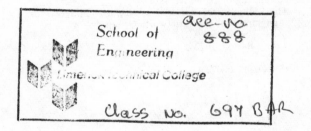
LONDON
NEWNES – BUTTERWORTHS

THE BUTTERWORTH GROUP

ENGLAND
Butterworth & Co (Publishers) Ltd
London: 88 Kingsway, WC2B 6AB

AUSTRALIA
Butterworth & Co (Australia) Ltd
Sydney: 586 Pacific Highway Chatswood, NSW 2067
Melbourne: 343 Little Collins Street, 3000
Brisbane: 240 Queen Street, 4000

CANADA
Butterworth & Co (Canada) Ltd
Toronto: 14 Curity Avenue, 374

NEW ZEALAND
Butterworth & Co (New Zealand) Ltd
Wellington: 26–28 Waring Taylor Street, 1

SOUTH AFRICA
Butterworth & Co (South Africa) (Pty) Ltd
Durban: 152–154 Gale Street

First published 1962 by George Newnes Ltd.
Second edition published 1968
Third edition published 1972 by Newnes – Butterworths,
an imprint of the Butterworth Group

ISBN 0 408 00072 4

*Phototypeset by Filmtype Services Limited,
Scarborough, England*

Printed in England by Fletcher & Son Ltd, Norwich

PREFACE

Before the publication of the first edition of this work in 1962, the average heating and ventilating designer and the students in the industry found the gathering of useful information on the subject of 'Cost Estimating' an extremely difficult and discouraging pursuit. The demand for the first and second editions particularly from students and the younger and more progressive engineers and technicians, architects, builders and surveyors has been most encouraging and the many letters I have received leave me in no doubt that the book has helped to fill a serious gap in heating and ventilating literature.

Over the years the book has become accepted by the building services, and building industries not only as a useful source of reference, but also as an acknowledged teaching aid.

This third edition has been completely revised to conform with the SI metric system of measurement, and decimal currency. Also, the various sections of the work have been extended as necessary to take account of the progressive changes in engineering and building practices.

Information on the compilation of 'Rates for Measured Work' is now given in Appendix F, and also in the Appendices, lists are included giving particulars of: Recommended publications of interest to the Estimator; Publications dealing with the Imperial/SI Metric changeover; Particulars of organisations from whom information useful to the Estimator is obtainable; Imperial/SI Metric and SI Metric/Imperial conversion factors and tables.

My sincere thanks are again due to:

THE INSTITUTION OF HEATING AND VENTILATING ENGINEERS for permission to quote from Section 17 of The I.H.V.E. Guide 1965, copies of which can be obtained from the Institution at 49 Cadogan Square, London S.W.1.

THE GAS COUNCIL, also the DEPARTMENT OF TRANSPORT AND POWER METEOROLOGICAL SERVICE, DUBLIN, for permission to publish the information contained in Tables E.1 and E.2.

THE BRITISH STANDARDS INSTITUTION for permission to quote from British Standard 350, Part 2, and from British Standard 1553, Part 4, copies of which can be obtained from the Institution at 2 Park Street, London W.1.

Southport   J. J. BARTON

PUBLISHER'S NOTE

Readers will appreciate that in a book concerned with an industry in which rates of labour and costs of materials are subject to change, it is impossible to ensure that those used are completely up-to-date. They may vary according to the particular area in which work is to be carried out, quite apart from increases agreed on a national scale between employers and trades unions. The worked examples should therefore be used as a guide to the method of calculation, substituting the rates applicable to any specific contract. These remarks also concern Governmental changes in S.E.T. and National Insurance contributions.

CONTENTS

1

THE PRINCIPLES AND ELEMENTS OF ESTIMATING

To introduce the subject, this chapter is devoted to some general observations on the principles and elements to be considered when preparing estimates.

An estimated cost of the proposed work is usually required for one of three main reasons:

1. To enable a tender to be prepared for submission by the contractor to the client or building owner direct, the tender being based on a scheme designed by the contractor, and using materials chosen by him.

2. As the basis of a tender, to be submitted in response to an invitation received by the contractor from a consulting engineer employed by the owner, or by the architect acting for the owner. In this case, the invitation to tender will be accompanied by a detailed specification, which will define the conditions to be observed, the quality, type, size, and other characteristics of the materials to be used, and also a set of plans, drawn to a scale of 1:100, giving a complete layout of the scheme. There will be a separate plan of each floor, including the roof space, and the basement. If the scheme is extensive, 1:20 scale drawings showing plans and elevations of the boiler room plant layout will be provided, in addition to a full set of the architect's elevations and sectional drawings.

3. Direct trading is not the only reason for estimating. The owner may be unwilling to commit himself to a project without suitable information as to its extent and cost. In this event a preliminary report and estimate is called for, and after considering these the owner decides whether or not to proceed.

The probable cost of works undertaken by Government Departments, Local Authorities, and other large organisations, must be known before the tender stage is reached, in order that the proposals may be examined by those Treasury officials, or Committees, on whose approval financial arrangements to enable the work to proceed will depend. Consulting engineers on the staffs of these Govern-

1

ment Departments and Local Authorities, or consultants specially engaged for particular projects, must therefore be able to forecast the result of future tendering with a reasonable degree of accuracy.

The estimator's job in competitive heating contracting is to build up a tender which if accepted will enable his firm to carry out the proposed work to the owner's satisfaction, and also make an adequate profit.

To tender is to *offer for acceptance* an estimate for a piece of work, or for goods, or services, the offer being referred to as *the tender*.

A number of items of cost go to the making up of an estimate for tendering purposes. Briefly, these are:

1. **Cost of materials, and cartage of materials to site.**
2. **Cost of labour to erect and install the materials.**

Up-to-date information upon the 'National Agreement on Wages and Conditions of Employment' is in publications available from the H.V.C.A. (See Appendix A.)

This main item will embrace several other elements, namely: (*a*) Depreciation of tools; (*b*) National Insurance and Redundancy Fund contributions by employer and Selective Employment Tax; (*c*) Annual Holiday Credits and Premiums for Sickness and Accident Insurance; (*d*) Lodging, and Daily Allowances; (*e*) Abnormal Conditions Payment; (*f*) Extra payment for work at high level, referred to as *Danger Money*; (*g*) Travelling Time and Payment of Workmen's Fares; (*h*) Merit Money; (*j*) Site Administration and Supervision.

3. **Overhead, or establishment charges.**
4. **Cost of Bond,** where this is called for.
5. **Design charge.** (To be included when the tender is based on a scheme designed in the contractor's office.)
6. **Contractor's Profit.**
7. **Prime Cost Items and Provisional Sums.**

A closer examination of these items will be helpful.

COST OF MATERIALS AND CARTAGE

Purchase of materials, plus the labour for erection, account for the major portion of the cost of any heating or ventilating installation.

The materials, including boilers, radiators, pipes, fittings, pipe hangers, brackets, supports, valves, fans, ductwork, and the many pieces of equipment required, are measured, or *taken off* the scheme drawings by the estimator, who then prepares a list of the measured materials. This list is referred to as a *List of Quantities*. When taking off is complete, the list of quantities can be priced and the items transferred to a cost sheet, which then becomes a *Bill of Quantities*. (See Chapter 5.)

Pricing of the general run of materials, such as boilers, pipes, valves, pipe fittings, etc., is done from the manufacturers' price lists, the figures in the Bill being the net prices, after deducting the manufacturers' discounts.

Items of equipment the prices of which may not be listed, such as pumps, centrifugal fans, and other machines and parts which have to be made to suit the needs of the particular scheme, are usually the subject of special quotations by the manufacturer, who will quote his net price for the equipment as specified.

Quality and type will be laid down in the specification, if prepared by a consultant, and all the invited tenderers will tender on the same basis.

If the tender is based on the contractor's own scheme he will put forward equipment and materials of his choice and will usually include, in addition to his tender and drawings, a Memorandum briefly but clearly explaining the scheme and naming the make and class of equipment and materials he proposes to use. In such a case the tender will usually be subject to the standard conditions of Trading issued by the Heating and Ventilating Contractors' Association.

Where a number of contractors are invited to submit schemes and tenders, each scheme, and also the materials to be used may vary, hence the need for the drawings and explanatory memorandum.

Cartage of Materials to Site

Materials which are in common demand, such as pipes, fittings, valves, etc., are usually kept in sufficient quantity at the contractor's stores; they are booked out from store and carried by the contractor's own vehicle to the site. This arrangement is followed by most contractors for jobs within a radius of say 25 to 30 miles (40 to 48 km) of their premises.

For distant contracts the materials are ordered carriage paid from the manufacturer or merchant, to be delivered on to the site, or to the nearest railway depot, in which case the railway will transmit to site.

Heavy and costly items such as boilers, automatic stokers, oil burners, large tanks, cylinders, radiators, pumps, and similar equipment, are usually delivered on site by the manufacturer, or for him by public transport, the cost being covered in the purchase price.

Late ordering and unplanned feeding of materials to the site are a common cause of labour wastage and subsequent loss of profit. Whether a contract has been obtained by competitive tender or not, labour kept idle for want of materials represents a direct loss, and can result in the job coming out on the wrong side.

Specials, prefabricated pipework, purpose-made supports for cylinders, pipe hangers, and all equipment required from specialist manufacturers must be on site when required if the contract is to run smoothly and profitably.

From the foregoing it will be appreciated that the charge in the estimate to cover the cost of cartage is an important and necessary item.

Reasonable expenditure for planned delivery of materials to the job should therefore be allowed for in the tender.

ESTIMATING LABOUR COSTS

Estimating the cost of labour is the most difficult part of making a tender and in the final analysis is largely a matter of experience.

So many variables can influence the cost of erection. All workmen are not equal in skill, and some work at a more leisurely pace than others; site conditions affect the speed of erection considerably. The rate at which work proceeds on a cold, wet site during winter is slower than for similar work carried out in the summer, and due allowance must be made for this when estimating labour charges. Work in partially completed buildings, especially multi-storey structures, becomes harder to execute during the months of December, January, and February, and during really severe spells may come to a standstill.

The pricing of operations during which the workman is exposed to the elements is most difficult. Concealed heating panel work where the panels have to be laid, levelled, welded, and hydraulically tested on the shuttering of floors under construction is a case in point.

Work carried out in subways and trenches, where conditions are often wet and foul, must receive special consideration.

Allowances have also to be made for work at high level, executed on scaffolds, in swings or cradles and from ladders and other awkward positions where erection of pipework becomes physically more difficult, and where the pace mentally becomes slower because greater control of movement must be exercised in the cause of safety.

Chapter 4 deals with labour estimating, and basic labour rates for the many operations involved in heating and ventilating work are listed. Suitable percentage increases are also suggested for work done under the abnormal conditions referred to above.

Site Administration and Supervision

The cost of site administration and supervision of the work on site is really a labour charge, and should properly be covered in the estimate, either as a percentage of the actual erection charge, or by estimating the actual time spent by the firm's travelling foreman on the particular site.

Some firms include this item in the overhead charges, but this method may not always cover the cost adequately, since certain types of work, especially when carried out under sub-contract conditions, call for more than average site planning and supervision.

A site foreman will usually be appointed by the firm to take complete charge of an extensive out of town contract, to be responsible direct to Head Office. This man's time will be completely taken up in supervising erection, planning progress, ordering materials, engaging local labour when necessary, and general site administration. The whole of his time and expenses must therefore be charged to the job.

A travelling foreman, whose time is divided between a number of sites, will be employed to administer and supervise the smaller contracts within a closer radius of the office. Except for brief visits to the office he is engaged entirely on outside work, and his time and expenses are charged to jobs under his control.

These men, who are usually ex-chargehand fitters or fitter-welders, must have had good experience of the trade, be tactful and patient, good at progressing the work, and be able to safeguard the firm's interest in dealing with the general contractor and with other trades. The foreman's skill in dealing with the main contractor's general foreman and sub-contractors on the site, has a considerable influence on productivity. Fitters can be kept idle for hours, often for days, waiting for pipe and radiator brackets to be built in; holes to be cut through walls and floors; pump and boiler bases; and many other operations, assistance for which the heating contractor, in his capacity of sub-contractor, will depend upon the general contractor.

Other aspects of site administration, such as day-to-day labour relations; ordering of materials for future delivery; setting out the more difficult sections of the installations and pipe runs; meeting the architect or consulting engineer to settle queries on points not clear in the specification and drawings, to discuss variations to the scheme, and to settle claims for extra work, all come within the scope of the foreman's duties, which, if competently carried out, will do much to ensure a profit at the end of the job and also enhance his firm's reputation.

When the heating contractor is the main contractor, as when carrying out installation work in existing buildings, he can exercise better control of progress. As the main contractor directly responsible to the owner or to the architect he is in a better position to arrange the timely execution of builder's, and other ancillary work. This ensures the heating labour being fully employed at all times, which is not only beneficial to the financial outcome of the current contract but releases the labour earlier for work elsewhere.

The cost of efficient site management and supervision is really a good investment, and must be included in the estimate.

Depreciation of Tools

Tools in poor condition, needing overhaul or replacement, are labour wasters which no contractor can afford. A good heating contractor makes a point of purchasing the very best tools, and provides facilities for their continual care and maintenance. Good tools in skilled hands are time savers. Worn and neglected equipment, especially the cutting tools such as screwing dies and hacksaw blades, pipe cutters, reamers, drills, and chisels, not only extend the time rate but also produce shoddy work.

Strongly-built lockers containing stocks and dies, chisels, files, reamers, hammers, drills, pipe cutters, hacksaws, electric drills, ratchet braces, chalk, centre punches, screwdrivers, taps for internal screwing, chain dogs, pipe wrenches, spanners, tape measures, jointing materials, etc., are usually provided by the heating contractor. These hand tools, together with benches, pipe bending machine, and portable forge, are delivered to the site and remain until the completion of the erection work, replacements being made as required.

The amount to include in the estimate for tool replacements is relatively small and can be covered by taking a small percentage of the net estimated labour charge for the job. The actual percentage will vary according to the care of tools exercised by the erection staff. A particular contractor will know from past experience of tools, what he must allow to cover this cost for any particular contract, and therefore no hard and fast rule can be stated. An allowance of 3 per cent of the net estimated labour charge will invariably cover this item. When heavy equipment, such as power-driven cranes, and special tools have to be hired to perform heavy or unusual operations, their charge will be a separate item in the Bill.

Quite a few other small items help to swell the labour charge. These are:

(*a*) The employer's share of *National Health, Social Security,* and *Redundancy Fund* contributions.

(*b*) *Overtime payment.* The normal working week is 40 hours, made up of 8 hours per day Monday to Friday inclusive, unless otherwise agreed between the employer and the operatives concerned. In cases where the employer and operatives agree, the working week may be reasonably extended, and overtime paid.

(*c*) *Lodging allowances* are paid when the operative is working on a job exceeding 30 miles (48 km) from his centre, or where it is not practical for the operative to travel daily to the job.

(*d*) When the job does not exceed 30 miles (48 km) from his centre, return daily travelling fares are paid to the operative; alternatively the employer must provide a suitable conveyance for the operative, to and from the job, each day. Where the operative is required to travel daily a distance exceeding 30 miles (48 km) from his centre to the job, he is paid a 'radius allowance'.

(*e*) When the operative is working at a distance not exceeding 30 miles (48 km) from his centre, and travels each day directly between his home and the job, allowances at agreed rates are paid by the contractor, provided the full normal hours are worked on the job.

(*f*) *Annual holidays with pay.* Every operative is entitled to two weeks' holiday with pay and one week of winter holiday. Each operative is credited by the employer with a weekly amount, agreed nationally by the employers' and operatives' organisations. This weekly payment is of course a charge on the contract upon which the operative is currently engaged, and must therefore be included in the estimated labour cost of the tender. A method of charging this cost when estimating for tendering purposes is explained in Chapter 5.

Working rules, and conditions for craftsmen, craft apprentices, and mates are the subject of national agreement between the Heating and Ventilating Contractors' Association, representing the contractors, and the National Union of Sheet Metal Workers, Coppersmiths, Heating and Domestic Engineers. The National Agreement covers minimum wage rates, extra payments for special conditions, and all matters affecting the operative's employment within the industry.

The estimator must have up-to-date knowledge of wage rates, and all payments and allowances which affect the labour estimate, and will keep a current copy of the National Agreement for this purpose. Full information on the National Agreement and other agreements affecting the industry is available in the *Heating and Ventilating*

Yearbook which is sponsored and compiled by the Heating and Ventilating Contractors Association, the Institution of Heating and Ventilating Engineers, the Heating and Ventilating Research Association, and the Hevac Association. (See Appendix A.)

OVERHEAD CHARGES

Overhead or establishment charges are usually included in the tender build-up as a percentage of the net cost of materials and labour. The charge covers the contractor's expenditure on director's fees, staff salaries and pensions scheme; rent; rates; insurances; light; heat; power for workshop and office machines; stationery; drawing office equipment; postage; telephones; depreciation of workshop machines and fittings, office furniture and machines; furnishings; office cleaning; floor coverings; advertising; travelling expenses; interest on loans, and bank charges. What the percentage shall be is obviously a decision for the firm concerned.

The annual expenditure on the items mentioned above will be related to the firm's annual trading turnover, usually by the firm's accountant, who will recommend an economic figure which may vary from 5 per cent to as high as 20 per cent.

The Head Office administration costs of two firms having nearly the same annual turnover may differ considerably. For example, one firm may own its central premises, purchased when property values were considerably lower than they are today, while the other may have to occupy highly rented offices and stores.

Advertising is another costly and sometimes essential item, especially for newly established firms. The older concerns who, by first-class work and sound trading methods over many years, have become well known and respected in the building industry, have not the same need to advertise as their younger and less well-established competitors.

The liquid financial position of the firm also affects overheads. Where bank loans are necessary the cost of such financial assistance becomes a charge on current trading.

It is thus apparent that the amount to include in the estimate to cover overhead charges cannot be stated as a general formula. It is, however, vitally important that the figure included in the estimate be based upon a true appreciation of the expenditure involved.

COST OF BOND

The contractor may be required to provide, at his own expense, a bond as security for due performance of the contract in an amount equal to 5 per cent of the contract sum. This is usually arranged with a bank or insurance company, who will of course make a charge for the service, the amount of which must be included in the estimate.

If the bond is arranged through the bank, the contractor will be charged interest on the value of the bond at the appropriate Bank Rate for the period of the bond, which is the contract period. An insurance company may charge approximately one quarter per cent on the value of the contract, for the period during which the bond is in force.

DESIGN CHARGE

This item is not included in estimates prepared for tenders based on outside designs, but should be taken into account when the contractor is invited to submit a complete scheme with his tender.

The item constitutes a labour and expenses charge only, the cost of paper, plan printing, etc., being covered under Overhead Charges.

If the scheme is for an existing building, commissioned directly by the owner or by his architect, a survey of the premises will be entailed; the time spent on this by the design engineer is charged to the job.

Time spent by a principal of the firm or by a senior design engineer in being briefed by the owner or by his architect may also be charged, together with travelling and cost of entertaining. Office time spent by principals, typing, and estimator's time in preparing the tender will usually be covered in overhead charges. An example of costing this item is given on page 10.

CONTRACTOR'S PROFIT

The elements which have been discussed so far make up the net cost of the installation, namely materials, labour, and overhead charges. To arrive at the selling price, the profit plus the value of all prime cost items and provisional sums must be added to the net cost, thus:

Net Cost = Materials + Labour + Overhead Charges.
Tender Price = Net Cost + Profit + P.C. Items + Provisional Sums.

The amount to be added for profit will be a percentage of the net cost, the actual percentage depending upon several factors.

When trade is depressed and work scarce, many firms engaged in competitive trade are forced to work on much lower margins than when trading conditions are good. In most firms the final decision as to profit will rest with a principal having intimate knowledge of the firm's financial position, state of the order book, future prospects, and planned future development of the business. He will also consider the type of contract involved, whether it is to be a direct contract or a sub-contract, the status and business character of the organisation or persons for whom the proposed work is to be carried out. The advice of the firm's accountant will be sought from time to time, especially when trade is good to ensure that present profits are designed to provide a sound basis for the future growth of the firm, and also to build reserves against a possible recession.

The allowance for profit may be any figure from $2\frac{1}{2}$ per cent to 20 per cent or more, and only the contractor, in the light of his needs and of the particular circumstances obtaining at the time of tendering, can decide.

Time *National Insurance Selective Employment Tax (S.E.T.) Expenses	Salaries: weekly rate at 40 h £	Salaries: hourly rate £	£	£
Senior Engineer, interview with owner, 8 hours away from office	48·00	1·20	9·60	
National Insurance (employer's contribution) and S.E.T.			0·43	
Expenses, travelling, lunch and entertaining			5·00	15·30
Design Engineer, 16 hours on survey (2 days)	33·00	0·825	13·20	
National Insurance and S.E.T.			0·86	
Expenses, travelling, and lunches			3·25	17·31
Design Engineer, preparation of specification and drawings, 176 hours	33·00	0·825	145·20	
National Insurance and S.E.T.			9·46	154·66
Tracer (female) 42 hours	15·00	0·375	15·75	
National Insurance and S.E.T.			1·47	17·32
Senior Engineer, supervision of design, 6 hours	48·00	1·20	7·20	
National Insurance and S.E.T.			0·32	7·52
Total to be added to labour item in Bill				£211·74

* Includes Redundancy Fund charge.

PRIME COST AND PROVISIONAL SUMS

A *Prime Cost*, or *P.C. Item* is the net amount included in the Specification and/or Bill of Quantities for equipment or articles to be provided (to the consulting engineer's or architect's approval), and paid for by the heating contractor, and to be handled, and if necessary fixed by him. (E.g. propeller fans, automatic stokers, oil burners.) The amount included for such items must be the net cost after deducting all trade and other discounts.

Provisional Sums may be divided into two distinct groups:

1. A Provisional Sum to cover the supply of a number of items (e.g. cooking equipment) to be selected by the engineer or architect, and handled by the heating contractor.

2. A Provisional Sum to provide for specialist work to be carried out by a sub-contractor (e.g. wiring to boiler room plant; thermal insulation).

Where P.C. or Provisional Sums are included in a heating specification for the supply only of materials or equipment, the amount of the P.C. or Provisional Sum should include a 5 per cent discount for the heating contractor.

Where P.C. or Provisional Sums are included to cover the supply and fixing of materials or equipment, such sums should include $2\frac{1}{2}$ per cent for the heating contractor.

P.C. and Provisional Sums for bringing public services to the building (gas, water, electricity) must be included in the main contractor's Bill of Quantities at net cost, and no discount added.

Equipment or work covered by Prime Cost Items is invariably priced by the owner or his consultant before tenders are invited. The heating contractor will include in his tender the cost of installing equipment, unless otherwise specified; he may add the cost of packing, carriage, and delivery to the site. An example showing how the Prime Cost is arrived at will make this clear. Suppose the consultant wishes to cover the supply of an oil burner in his specification, the list price of which is £300. The Prime Cost to be included in the specification will be determined thus:

	£
List price of oil burner and controls	300·00
Less 25 per cent trade discount	75·00
Net cost	225·00
Add 1/19th to cover contractor's discount	11·84
Carried forward	236·84

Amount of P.C. Item which includes 5 per cent for
   heating contractor   236·84
Cost of packing, carriage and delivery   5·75

Total cost   £242·59

The provision and installation of such an oil burner might be covered by a Provisional Sum, in which case the heating contractor would be entitled to the appropriate $2\frac{1}{2}$ per cent discount. The cost of the installation work will include erection and fitting to the boiler, together with the wiring of all electrical controls from an isolating switchfuse in the boiler room. If the Specification does not call for installation by the manufacturer, the heating contractor may, if he employs competent workmen, include this cost in his tender.

Generally speaking Prime Cost Items serve a useful purpose only where there are certain items specified that can vary widely in quality and price.

In this way all tenders are kept on the same basis, and the client's choice of equipment is safeguarded.

The Contingency Item

Reasonable provision must also be made for minor additions and alterations to the scheme which may become necessary after the contract is let, and which could not be foreseen during the design stage. This is called the *Contingency Item* and is included in the Specification as a Provisional Sum. The amount of the Contingency Item will be a small percentage of the estimated total cost of the work, usually 5 per cent for jobs up to £1 000 in value, falling to $1\frac{1}{2}$ per cent up to £10 000, and 1 per cent when the contract exceeds £10 000. The Contingency Sum will be expended during the course of the contract as directed by the owner or his representative and the amount not expended at the end of the contract will be deducted from the final total.

Provision for Testing the Installation

Before the installation can be handed over to the owner full thermal and circulation tests are necessary, and the cost of labour, fuel and use of instruments should be included in the tender.

Where the heating installation is a sub-contract the cost of fuel for testing is often included in the main contractor's Bill, in which case the heating contractor will provide labour and instruments only. The cost of instruments will normally be covered by the allowance for tools, discussed on page 6. The time taken to test and balance an installation will depend upon its size; labour rates for this item are included in Chapter 4.

If a supply of fuel for testing is not included in the main contractor's Bill, it may appear as a P.C. Item in the heating specification, or alternatively the owner may supply direct. Unless clearly instructed to do so in the Specification, or by letter or official order, the heating contractor should not include the cost of fuel in his tender.

Electricity for operating pumps, and boiler firing equipment during tests, is supplied by the owner and no allowance for this is necessary in the heating tender. Initial supplies of lubricating oil for machine gearboxes and bearings are the responsibility of the contractor, and the cost, together with that of other sundries such as jointing materials and oil for lubricating pipe threading operations must be covered in the tender.

Other charges such as official fees to Water Boards for stamping fittings, taps, and valves; service charges to water, gas, and electricity authorities have to be met by the heating contractor when specified, and the cost included in the tender.

Variations to Contract

In the heating and ventilating industry tenders are submitted on a lump sum basis, prepared from adequate drawings and specifications, and in accordance with the *Standard Method of Measurement, Section XVI*, the successful tenderer is required to prepare a priced Bill of Quantities, upon which subsequent variations to the contract are priced. Copies of *Standard Method of Measurement, Section XVI*, can be obtained by recognised members of the industry from the publishers, The Royal Institution of Chartered Surveyors, 12 Great George Street, London S.W.1, and the National Federation of Building Trades Employers, 82 Cavendish Street, London W.1.

Where the successful tenderer is not asked to submit a priced Bill of Quantities he will be required, as an alternative, to provide a Schedule of Prices for 'work as fixed', upon which the cost of variations to the scheme can be priced. Usually, the contractor completes the priced schedule after being advised that he is the successful tenderer, subject to the owner's approval of the completed schedule. The cost of preparing the schedule is normally regarded as an overhead charge.

Additional works not subject to a separate tender and therefore not included in the 'Bill of Quantities' or a 'Schedule of Prices for

Costing Variations', will be charged for at the cost of those additional works at the time the works are carried out, plus percentage additions to Labour, Materials, Fares, Allowances, and Cartage. The percentage additions for such daywork items (and also for hourly jobbing work rates) are then negotiated between Contractor and Client on an individual basis.

The Heating Contractor will base his tender upon the work being done during the normal working hours set out in the 'National Agreement as to Working Rules and Conditions in the Industry'.

Overtime authorised by the owner, or his agent (the consulting engineer or the architect) is therefore an additional charge, to be paid at the rates in force at the date when the work is executed.

The system as installed finally may, due to alterations decided upon during the course of the contract, be somewhat different from the scheme as designed. In order that the owner may have a true record of the scheme as installed, the contractor, when specified, will include in his tender for providing a full set of *as installed drawings* on linen.

Conditions of Tender. A standard form of conditions of tender for use in connection with mechanical engineering systems in buildings, is published by the A.I.H. Committee, on behalf of The Association of Consulting Engineers, Abbey House, Victoria Street, London S.W.1, The Institution of Heating and Ventilating Engineers, 49 Cadogan Square, London S.W.7, and The Heating and Ventilating Contractors Association, Coastal Chambers, 172 Buckingham Palace Road, London S.W.1, from whom copies may be purchased. (See Appendix A.)

Guarantee of Results

When the installation is the subject of a direct contract between owner and heating contractor, the latter will, if required to do so by the owner or his architect, provide a Guarantee of Results. This will embrace materials, workmanship, performance of the system as to temperature, air quantities, etc. The owner may also require some authoritative information from the contractor upon annual costs. Materials are adequately guaranteed by all reputable manufacturers, and the contractor will of course be prepared to guarantee the work of his own labour.

The owner may ask for a *defects liability period* of 12 months.

The guarantee of performance covers:

    1. Temperatures to be maintained in rooms, with specified

outside conditions, when the boiler plant is working at specified temperatures. (The heating engineer should only give a temperature guarantee under defined operating and usage conditions, which, he must clearly state. Complaints regarding lack of heat in buildings fitted with fully adequate heating systems very often come from people who insist on most of the windows and doors in the building being wide open. These people expect a performance with which the sun alone could cope.)

2. Fuel consumption during a heating season of specified length, and average weather, using a specified fuel.

3. Power consumption of the equipment installed under the contract.

For ventilation and plenum heating, the guarantee covers room air temperatures, and air quantity delivered to each room and, when specially called for, noise levels in rooms when plant is operating.

Estimation of Running Costs

If the owner wishes to have information as to the probable running costs of the proposed installation, the contractor's memorandum accompanying the tender should include a full report, giving the estimated running costs for stated periods of operation. The period for heating is usually taken as 210 days, with agreed extensions for hospitals, old people's hostels, and similar buildings. Hot water costs will usually be based on a full year, the weekly consumption varying according to the use to which the building is put. For example a hospital requires a full hot water service on 365 days per year, while an office block operating on a five-day week has an annual usage period of 261 days.

Plenum heating and air conditioning may be operating either as heating and ventilating systems, or solely for ventilating and cooling, during periods which normally coincide with the seasons.

The contractor should take special care when submitting such reports to state carefully and exactly the periods, operating and weather conditions, rate of air change, and also the values of all factors involved. Furthermore, he should give his methods of calculation, and show precisely how he has arrived at the result. The memorandum should also make clear to the owner that the results put forward are estimated and therefore cannot be included in any guarantee.

When a consulting engineer is employed to prepare complete plans and specifications, it becomes his responsibility to provide any guarantee as to the performance of an installation properly carried out in accordance with his own scheme, and here again the guarantee must not be too liberal, and should only be given under conditions laid down by the engineer.

The consultant will also provide a report on estimated running costs, as a special part of a much wider report giving the economic and engineering characteristics of his proposals.

Investigation of Alternative Systems

Very often the owner will require his consulting engineer to investigate some special form of equipment which he, the owner, may have in mind.

During the early stages of a large building project the consulting engineer may be required to give an opinion on the advantages and disadvantages of several methods of heating and ventilating as applied to the particular project. The report will contain detailed descriptions of all plant and equipment, comparative capital and running costs, and also detailed estimates of the cost of the structural work for each method considered. Included in the latter, in addition to the usual builder's work, will be the estimated capital cost of boiler house, chimney, fuel storage, and cost of access or of roads which are specially required for fuel delivery. A complete and accurate picture of the engineering and economic characteristics of each method is presented. The cost of access and roads is mentioned because acceptance of a particular fuel, where there is restricted access to the boiler house, may necessitate the owner purchasing land to ease the position. When this is necessary the cost must be included in the engineer's capital estimate.

Approximate Estimates

It is often necessary to give approximate estimates at short notice. An architect may require a quick engineering price to include in a proposal for consideration by his client. A *price record book* is most useful for this purpose. The book should contain particulars of all jobs which the firm have completed, including such details as type of building, purpose of building, cubic capacity of building, calculated heat losses of building, type of installation, tender figure, profit made, cost of various portions of the installation as installed, final cost of job, and of course the date of installation. For schemes of similar type and of the same specification to be installed in buildings of similar usage, exposure, and construction, the record book provides a reliable means of producing an approximate price. It is of great value both to the consulting engineer and to the contracting engineer, and its use is discussed in more detail in Chapter 5.

Builder's Work

The main elements affecting the engineering portion of the tender have now been reviewed.

Other work, such as building and plumbing operations, is necessary for the functioning of the heating installation, and the cost must be accounted for, either in the heating contractor's tender, or as Provisional Sums in the main contractor's tender.

If the invitation to tender comes from a consulting engineer he will be responsible for preparing a Schedule of Builder's Work, which is then priced by the architect or by the quantity surveyor, who will arrange for its inclusion in the main contractor's Bill of Quantities.

Where the heating system is intended for an existing building, the consulting engineer prepares the builder's work schedule, and includes a Provisional Sum in his specification to cover the cost. The heating contractor, who in this case becomes the main contractor, sub-lets the work to a builder.

When the heating contractor is invited by the building owner, or his architect, to submit a scheme and tender for installing heating in an existing building, the builder's work schedule is prepared by the heating contractor. He usually sends the schedule to a builder for pricing, and includes the net cost, plus profit, in his tender.

Costing

Sound costing is essential for successful contracting and jobbing work. It is true to say that many firms, especially the smaller ones, fail because of inadequate job costing. Efficient costing not only assists the business by giving a weekly statement as to the financial state of each contract on the books, but is also of immense value to the estimator, especially in assessing labour rates.

A weekly check should be kept on all items affecting the actual cost of the work, and the total expenditure on work done to date should be compared with the amount included for that work in the tender. In this way tendencies which often lead to serious loss if allowed to continue, can be remedied in time, and the profit margin maintained. Careful costing will reveal much valuable information regarding the quality of labour employed by the firm. This is because the basis of the costing of the labour item is the simple time sheet.

The terms of most contracts provide for monthly payments to

the contractor up to the value of 90 per cent of work completed, and of materials safely on site. Application for this interim payment is made by the heating contractor to the owner each month. If the job is being properly costed, up-to-date cost sheets are always available upon which the contractor can assess the amount of his monthly claim, and which can be examined by the owner's architect or consulting engineer, who may wish to check the validity of the amount claimed.

The contractor should take care to choose a simple costing system which can be operated by any member of the office staff. The heating contractor's business does not, by its practical nature, require an intricate system, which may be costly to operate and add unduly to the cost of overheads. Examples of costing are dealt with in Chapter 6.

The unique position of the properly trained Site Foreman or Supervisor, not only in improving site productivity, but also in effective cost recording, is not always appreciated.

A Site Supervisor who, in addition to his site management duties is made sufficiently aware of the design, accounting, and managerial problems, associated with his firm's contracting business, fulfils a very important role in the industry.

2

SPECIFICATION AND DRAWINGS

The specification as applied to heating and ventilating work consists of a detailed description of the work and materials to be used. It also includes full instructions to the contractor as to the method and order in which the work is to be carried out, and clarifies any special terms and conditions necessary for the proper fulfilment of the contract.

In conjunction with the working drawings, the specification should provide a clear and concise picture of the proposed scheme for both estimating and installation purposes.

Floor plans indicating pipe runs and radiator positions are usually drawn to a scale of 1:100. For very large schemes where several buildings may be heated from a central boiler house, smaller scale plans, 1:200, may be used to show the main runs between the boiler house and the buildings. Detailed drawings of the boiler room plant and other intricate piping and plant arrangements which would be difficult to portray in small scale drawings are drawn to a scale of 1, 20, or 10 when the size of the paper will allow it.

Except on the larger scale detail drawings, pipes are indicated by single lines as follows:

Pipes in roof or above ceiling.

Pipes at floor level.

Pipes at high level in the room.

Pipes under floors, or in trenches.

The direction of flow in pipe runs is usually shown by arrow heads on the pipe thus

or by small arrows drawn parallel to the pipe, thus

The abbreviations and graphical symbols more generally used in preparing heating plans are shown in Table 1. B.S. 1553: Part 4: 1956, and 1553: Part 1: 1949, give full details of all recommended symbols.

Figure 1 shows a low pressure gravity hot water heating installation in a large church hall. The scheme is indicated entirely by lines, symbols, and abbreviated instructions, which are easily understood by the estimator and by the installation fitter. Certain portions of

19

Table 1. GRAPHICAL SYMBOLS GENERALLY EMPLOYED FOR INDI-
CATING HEATING WATER SUPPLY AND VENTILATION ON WORKING
DRAWINGS AND DIAGRAMS

DESCRIPTION	SYMBOL
PIPE AT LOW LEVEL	
PIPE AT HIGH LEVEL	
PIPE BELOW FLOOR LEVEL	
PIPES CROSSING BUT NOT CONNECTED	
DIRECTION OF FLOW IN PIPE	
DIRECTION OF FLOW IN PIPE	
VERTICAL PIPE RISING TO ABOVE	T.A.
VERTICAL PIPE DROPPING TO BELOW	T.B.
NON−RETURN VALVE	
STOPVALVE, OR STOPCOCK	
RADIATOR ON SINGLE PIPE CIRCUIT	
RADIATOR ON TWO PIPE CIRCUIT	

Table 1 (contd.) GRAPHICAL SYMBOLS

DESCRIPTION	SYMBOL
AXIAL FLOW FAN	
CENTRIFUGAL FAN	
UNIT HEATER	

THE FOLLOWING ABBREVIATIONS ARE COMMONLY USED
FOR IDENTIFICATION AND DIRECTIVE PURPOSES
ON HEATING AND WATER DRAWINGS

FLOW PIPE	F.	LOW LEVEL	L.L.
RETURN PIPE	R.	HOT WATER SUPPLY	H.W.S.
HOT WATER SPACE HEATING	H.W.H.	COLD WATER	C.W.
FLOW & RETURN PIPES	F.R.	RADIATOR	RAD.
PIPE RISING TO ABOVE	T.A.	LAVATORY BASIN	L.B.
PIPE DROPPING TO BELOW	T.B.	AIR COCK	A.C.
HIGH LEVEL	H.L.		

RADIATOR CONNECTIONS

TOP AND BOTTOM OPPOSITE ENDS	T.B.O.E.
TOP AND BOTTOM SAME END	T.B.S.E.
BOTTOM OPPOSITE ENDS	B.O.E.

Table 1 (contd.) GRAPHICAL SYMBOLS

DESCRIPTION	SYMBOL
TOWEL RAIL	
HOT OR COLD WATER DRAW-OFF POINT	
MIXING VALVE	
BOILER	
DIRECT HOT WATER STORAGE CYLINDER	
INDIRECT HOT WATER CYLINDER OR CALORIFIER	
COLD WATER STORAGE TANK OR FEED AND EXPANSION TANK	
CENTRIFUGAL PUMP	

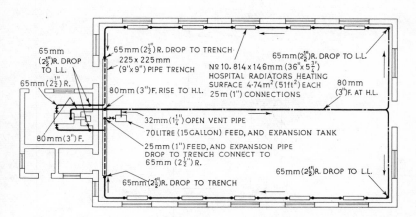

Figure 1. Plan of low pressure gravity hot water heating in a church hall (Scale 1:100)

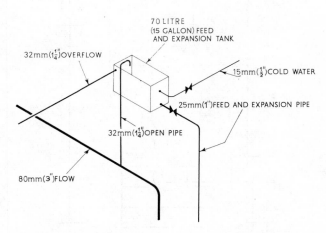

Figure 2. Detail of feed and expansion tank connections (Not to scale)

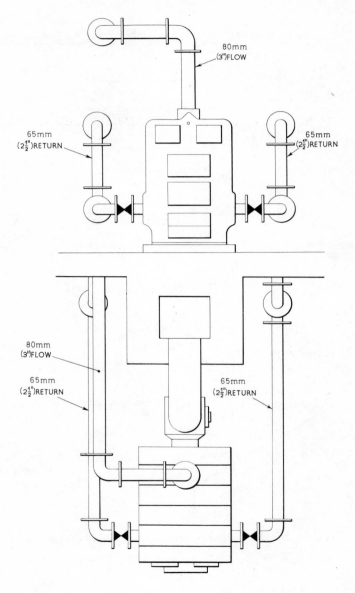

Figure 3. 1:20 details of boiler room plant (Not to scale)

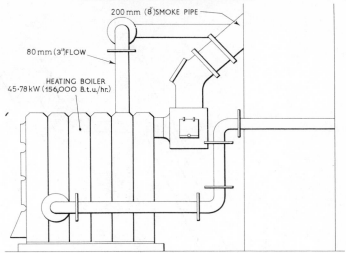

200 mm (8″)SMOKE PIPE

80 mm (3″)FLOW

HEATING BOILER
45·78 kW (156,000 B.t.u./hr.)

Figure 3 (contd.) Side elevation of boiler room plant

pipework which may not be clearly defined on the plan are often detailed with small isometric single line sketches, similar to Figure 2, which shows the feed and expansion tank and its associated pipework.

The boiler room plant is shown in 1 : 20 detail (Figure 3).

For estimating purposes an isometric sketch of the scheme, drawn to scale, is most useful.

Figure 4 is an isometric diagram of the scheme shown in plan in Figure 1, and from this all pipes, fittings, boiler, radiators and all other equipment may be measured and listed for estimating.

Standard Specification

A form of standard specification for heating work is often used by large Local Authorities and other bodies responsible for extensive capital works programmes. The use of a standard specification greatly reduces the work on individual schemes. Large numbers of the standard specification are printed; a copy is sent out with each enquiry, in addition to the scheme drawing, together with a supplementary specification containing particulars of materials and other items peculiar to the particular scheme.

A typical standard specification applying to heating, water, steam boiler installations and ventilation is given in the following pages.

The Ministry Of Public Building And Works, *Standard Specification (M amd E) No. 3: 1968*, provides an excellent guide to the composition and range of the standard form of Specification.

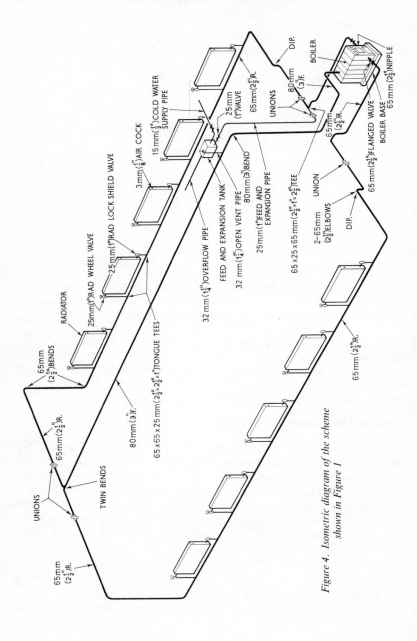

Figure 4. Isometric diagram of the scheme shown in Figure 1

TYPICAL STANDARD SPECIFICATION AND REGULATIONS APPLYING TO ENGINEERING WORK IN CONNECTION WITH HEATING, WATER, STEAM INSTALLATIONS, VENTILATION AND FUEL BURNING EQUIPMENT

Section A

Extent of work

1. The engineering contractor, hereinafter referred to as *The Contractor*, must include in his tender for the supply, delivery, storage and installation of all materials and equipment necessary to complete the installation, in accordance with this standard specification, the drawings and any supplementary specification issued by the engineer. The term *Contractor* shall also be taken to mean Sub-Contractor where this term applies. The expression *Engineer* shall be taken to mean the consulting engineer responsible for the design and installation.

Visiting the site

2. The Contractor must visit the site to ensure his being conversant with local conditions as on no account will any extras be permitted due to his failure to do so.

Where systems are to be installed within existing buildings, the Contractor shall, before tendering, examine the scheme as shown on the drawings and as specified, in relation to the building or buildings, and shall then be deemed to have a good knowledge of the scheme or schemes, and claims for costs for extra labour and/or materials will not be allowed when it can be shown that this examination has not been closely carried out.

Method of tendering

3. In submitting his tender the Contractor shall detail the prices in the manner shown below.

(1) Heating system.
(2) Hot water supply system executed in copper.
(3) Hot water supply system executed in galvanised steel.
(4) Cold water supply system executed in copper.
(5) Cold water supply system executed in galvanised steel.

    (6) Cooking equipment.
    (7) Gas pipework from meter to gas-using equipment.
    (8) Gas service.
    (9) Water service.
  (10) Ventilating work.
  (11) Provisional Sums.
  (12) Prime Cost Items.

Plans and specifications

4. Plans showing the required layout of the proposed installations are sent out together with a Supplementary Specification where deemed necessary, and the Contractor must include for all labour, supervision, equipment and materials for carrying out the work as defined by the plans, by this standard specification and by any supplementary specification.

Where individual fittings and small materials are necessary, the exact nature and shape of which cannot be clearly defined in the tender drawings, the Contractor is assumed to have included in his tender such as will conform to good modern practice.

The plans and Supplementary Specification must be returned to the Engineer separately from the tender.

Samples

5. The Contractor must submit samples of all materials and obtain the approval of the Engineer before using them.

Workmanship and measurement of work

6. All materials and workmanship throughout the entire Contract to be new and the best of their respective kinds and to the complete satisfaction of the Engineer, acting for the Owner, under whose supervision the work will be carried out. All materials must be of approved manufacture.

The Contractor is to take his own measurements on the site and is to be entirely responsible for the quantities required.

After the work is completed, including any additions or deductions, it is to be measured by the Contractor in the presence of the Engineer.

Local bye-laws and regulations

7. The Contractor shall ascertain and act in accordance with any local bye-laws and regulations which may affect the work under this Contract. Any cost for the stamping of valves, fittings and the like must be covered by the Contractor in his tender.

Reception and storage of materials

8. The Contractor shall provide proper storage for all materials, tubes, fittings, etc. delivered to the site, which must be received at the site by the Contractor. Tubes and fittings stored on the ground outside will not be approved for use. The Contractor will be fully responsible for reception and storage. Where tubes, radiators, boiler sections, etc. are stored outside, proper sheeted racks and other supports must be provided.

Liability for defects

9. The Contractor is to make good any defects in his work which may be due to faulty workmanship or materials supplied by him, which may arise within 12 months from the date of acceptance of the completed installation by the Owner. This applies to all materials, equipment, and completed work.

Variations to scheme and prices

10. The Contractor shall quote a lump sum tender for the work as detailed in the specifications and as shown on the drawings, and is to fill in the variation schedule attached to the form of tender, giving the cost of pipes, etc. supplied and fixed, as included in the Main Contract.

Should the Contractor prefer, he may submit a priced Bill of Quantities based upon the specifications and drawings, in accordance with the Standard Method of Measurement, Section XVI, upon which variations to the Contract shall be priced. If the Contractor wishes to submit a priced Bill of Quantities, he must indicate this in writing when submitting his tender.

Any variation, both in cost of materials and general scheme, will be based on these figures.

Tests

11. The following pressure tests must be carried out by the Contractor before or during the progress of the work as determined by the nature of the tests:

Panel Heating. After manufacture each panel coil must be tested to an air pressure of not less than 3500 kN/m² (35 bars) while the panel coil is immersed in clear water. Before the air pressure test mentioned above is carried out, an 11 mm diameter steel ball must be blown through each panel. A test certificate certifying that these two tests have been carried out must be provided for each panel. After each panel has been laid in position on the shuttering or suspended ceiling, and before the concrete is cast, a hydraulic test at 2100 kN/m² (21 bars) must be applied for four hours in the presence of the Engineer or the Clerk of Works. Mains and risers to be tested to 1½ times the working pressure for 30 minutes.

Plaster. The Contractor must include in his tender for one visit of the Invisible Panel Warming Association's Plastering Inspector, and due notice of his impending visit must be given to the Engineer in writing.

Manufacturers test certificates to comply with the appropriate British Standard shall be provided by the contractor for all plant and equipment to be operated under pressure.

The I.H.V.E. Commissioning and Circulation Test (I.H.V.E. Guide 1970, Section B17) must be carried out before the installation is handed over to the Owner.

Where an item of plant is not covered by a British Standard, it shall be hydraulically tested to 1½ times the operating pressure, or 1400 kN/m² (14 bars) whichever is the greater.

Every complete hot water space heating installation, or domestic hot water and cold water system shall be pressure tested in accordance with C.P. 341.300, Section 0·601, and C.P. 310 respectively.

Drawings sent in for approval

12. Drawings shall be sent in to the Engineer as required for general approval but approval of such drawings will not exempt the Contractor from his liabilities under the Specification.

Record drawings

13. On completion of the works the Contractor must supply one complete set of coloured linen drawings showing all the services covered by the Contract. All equipment, panels, radiators, pipes, taps, valves, etc. to be shown in their correct positions together with one set of 1 : 20 detail drawings of the boiler room plant, including automatic stokers or oil burners and thermostatic controls, cooking, ventilating and all plant included in the Contract or Sub-Contract.

Builder's work

14. All builder's work incidental to the installation of the systems specified will be carried out by the appointed Building Contractor but the Engineering Contractor is to supply the Builder with details of boiler and pump bases and of all cutting away, and will be responsible for all marking out, etc. Any extra cost incurred for builder's work due to faulty marking out or failure of the Contractor to give proper instructions to the Builder at the time required for the proper progress of the work shall be borne by the Contractor.

Where a Builder is not employed by the Owner, as in those cases where the Engineering Contractor is the Main Contractor, the Engineering Contractor must include in his tender and be responsible for the carrying out of all builder's work necessary to complete the installation to the complete satisfaction of the Engineer. Painting work normally included by the Builder, and all similar work, is covered by this clause.

Panel installation

15. The invisible panel section of the heating system must be provided and installed only by a member of the Invisible Panel Warming Association who has carried out a number of invisible panel warming installations, and firms tendering who are not members of this Association must provide for this in their tender. For this purpose the panel section of the work shall be taken to include all panels, panel connections, and all flow and return risers from the basement mains and all venting and emptying arrangements.

Delivery forecast

16. When submitting his tender the Contractor shall give the manufacturers' delivery forecast for all materials.

Sleeves

17. All pipes where they pass through walls and floors of the building must be fitted with telescopic galvanised special pipe sleeves or flush pipe sleeves, and wall or floor plates as specified in the Supplementary Specification.

Where pipes are exposed the sleeves should have thimble ends, and when hidden by casings or otherwise obscured, the sleeves should have plain ends.

Badly fitted sleeves which, due to carelessness by the Contractor or Builder, are left proud of the finished wall or floor or are otherwise unsightly, must be renewed. It is the responsibility of the Contractor to see that sleeves are properly and neatly fitted, and the Contractor must bear all extra cost involved in remedying bad work.

Open ends

18. During the progress of the work open ends of pipes, tees, brackets, etc. must be closed with metal plugs. Plugs made from paper or other materials will not be allowed. Work not protected in accordance with this clause will not be accepted.

Pipe burr

19. Suitable reamers must be provided for completely removing the inside burr from steel and copper pipes. The Contractor should understand that if evidence of unmoved burr is found in one small portion of the work during installation, the Engineer will require the remaining work to be dismantled for his inspection.

Tubes

20. Tubes shall comply in all respects with the appropriate British Standard as follows, and as indicated in the supplementary Specification.

Heating. Low temperature hot water, B.S. 1387 mild steel ungalvanised medium grade.

Concealed panel systems. Panels. B.S. 1387 mild steel ungalvanised heavy grade.

Medium and high temperature pressurised systems. B.S. 1387 ungalvanised heavy grade.

Steam heating systems. Steam pipes. B.S. 1387 ungalvanised heavy grade. Condensate pipes. B.S. 1387 ungalvanised heavy grade. Or in copper. B.S. 2871: Part 2.

Domestic hot water systems. B.S. 1387 galvanised medium grade. Or copper. B.S. 2871: Part 1.

Cold Water Services. B.S. 1387 galvanised medium grade mild steel inside the building, and B.S. 1387 galvanised heavy grade for buried pipework. Where copper tubes are specified for cold water services they shall be B.S. 2871: Part 1 for surface work and for buried pipework. Where lead or plastic pipes are to be used particulars of the material will be given in the supplementary Specification.

Where dissimilar metals are used the Contractor must provide and fit approved insulating joints to prevent corrosion due to electrolytic action.

Pipe fittings and joints

21. Fittings shall be screwed malleable for all sizes up to and including 65 mm ($2\frac{1}{2}$ in) and flanged for 80 mm (3 in) work and over. Should copper be chosen for pipework systems, the type of fitting required will be indicated in the Supplementary Specification or on the drawing. Generally, for copper water systems, capillary soldered joints or compression fittings of approved manufacture should be used, as specified in the Supplementary Specification.

Failures have occurred on some installations using compression and capillary fittings and investigation has shown the failures to be mostly caused by poor workmanship. The attention of the Contractor is particularly drawn to ensuring that the fitter should cut pipes of correct length to ensure the pipe end completely filling the recess in the fitting.

During the progress of the work sections of the pipework will, after erection, be partially dismantled to enable the Engineer to examine the joints, and where pipes are not of full length, the section of work affected will not be accepted. If capillary fittings are used they must be installed by workmen with past experience of such fittings.

For screwed work the Contractor is expected to examine the fitter's pipe thread dies at frequent intervals and to ensure that these tools are kept in constant first class condition. Threads must be correctly and cleanly cut and must be the right length for each pipe size. Threads must be cut to engage fully with the fitting and must not be too long nor too short. All screwed joints must be made with best quality hemp and graphite compound. Flanged joints must be machine faced and ground or made with graphited asbestos jointing gaskets or rings. Some welding of joints on heating mains in trenches and in the boiler room, pump room and other fully accessible positions will be allowed where this method will benefit the installation without jeopardising the easy renewal of sections of the pipework during future maintenance work. Welded joints will be allowed when indicated in the Supplementary Specification.

Valves

22. Valves for heating and hot water and, where specified, for cold water, shall be of approved manufacture; to be full-way gate pattern of gunmetal throughout up to and including 65 mm ($2\frac{1}{2}$ in) diameter, and cast iron for 80 mm (3 in) diameter and over, with screwed ends up to and including 65 mm ($2\frac{1}{2}$ in) diameter and flanged for pipes of 80 mm (3 in) diameter and over.

Cast iron fullway valves shall be fitted with renewable gunmetal seating rings. Valves fixed in rooms shall be easyclean pattern. Radiator valves shall be of approved manufacture. Keys and dust caps must be provided for all lock-shield valves.

For steam, condensate, and for cold water unless otherwise specified, gunmetal globe pattern valves shall be used, fitted with appropriate valve discs and seats.

Brackets and supports

23. Pipe lines to be securely fixed by means of L.C.C. type built-in schoolboard brackets and to be black malleable in all positions excepting the boiler house, where all brackets used shall be galvanised malleable.

Where standard brackets cannot be used, e.g. in the boiler room and pump room, purpose-made brackets, hangers, or supports must be used. The design of purpose-made brackets must be approved by the Engineer before use.

The maximum spacing of pipe supports shall be in accordance with the following table.

Nominal bore of pipe		Steel pipes		Pipe size B.S.2871 (o.d.)		Light gauge copper pipes	
		Spacing horizontal runs	Spacing vertical runs			Spacing horizontal runs	Spacing vertical runs
mm	in	metres	metres	mm	in	metres	metres
15	½	1·8	2·4	15	½	1·2	1·8
20	¾	2·4	3·0	22	¾	1·8	2·4
25	1	2·4	3·0	28	1	1·8	2·4
32	1¼	2·7	3·0	35	1¼	2·4	3·0
40	1½	3·0	3·7	42	1½	2·4	3·0
50	2	3·0	3·7	54	2	2·4	3·0
65	2½	3·7	4·6	63	2½	3·0	3·7
80	3	3·7	4·6	76	3	3·0	3·7
100	4	4·0	4·6	108	4	3·0	3·7
125	5	4·6	5·5	133	5	3·7	3·7
150	6	5·5	5·5	159	6	4·6	3·7

Labelling

24. All valves in the boiler room, pump room, roof spaces and trenches must be clearly labelled with approved pattern non-ferrous metal plates with plain stamped lettering.

Mains and other pipework

25. Provide and fix all steam, condensate, hot water heating, hot water supply, cold water, gas and air mains and other pipework as shown on the drawings. Long radius elbows should be used in preference to bends. Square elbows will not be allowed. Pipe runs must follow the line of the building and sets must be made around all piers and other projections. Pipes must also be set to follow the lines of all recesses. Where details have not been approved the Contractor must discuss all pipe runs with the Engineer and obtain permission before proceeding with the installation. Provision for proper drainage for all runs must be made. Special attention must be paid to the installation of vertical pipes which must, where possible, be run in chases provided by the Builder. Careful arranging of pipework, especially where more than one service occurs or occupies the same trench or chase, will be insisted upon to achieve the neatest appearance or preferably to conceal pipework not required as heating surface. Workmen employed must be fully skilled and capable of executing a really first class installation. In this connection clean joints will be insisted upon and tool marks or imprints on valves and fittings will not be tolerated. Spanners of the correct sizes must be issued to the fitters and used by them, and stillson pipe wrenches and similar tools must be confined to the use for which they are intended.

Draw-off taps

26. The Contractor will provide and fix all hot and cold taps to draw-offs except those actually secured to or serving sanitary, kitchen or other fittings, when he shall run his pipes to within an average distance of 1 m from the actual connection to each fitting, and the Builder's plumber will connect to the fitting.

Air venting

27. The Contractor must allow in his tender for the proper air venting of all steam, condensate, heating, hot and cold water pipes, and appliances, whether provision for this is shown on the drawings or not.

Open vents, automatic air vents, air bottles and wheel-operated air vents will be allowed as approved by the Engineer. Loose key air cocks or valves will not be allowed on pipes, radiators or other equipment unless otherwise specified.

All open pipes must terminate over a cold water tank or expansion tank with a double bend.

Drainage

28. The Contractor must allow in his tender for the proper drainage of all mains and branches, storage and pressure vessels, boilers, steam, condensate, heating, hot water, cold water, and gas installations, whether provision for this is shown on the drawings or not.

Provision for drainage shall consist of 10 mm ($\frac{3}{8}$ in) gland cocks or other approved valves, every cock or valve to be fitted with a hose connection.

Stopcocks, hot and cold water and gas

29. To each draw-off or pipe serving a fitting or series of

fittings, provide and fix an approved screw down easyclean
stopcock with pinned jumper.

Radiators

30. In the positions shown on the drawing, provide and fix
cast iron radiators of approved make. All radiators to be
wall or floor type as required and to be fitted with a wood
or plastic wheel easyclean valve on the inlet, and a lockshield
easyclean valve on the return. Each wall type radiator to be
fixed on built-in cast iron or steel radiator brackets and
supported with a built-in 10 mm ($\frac{3}{8}$ in) back stay. Provide
and fix to each radiator 3 mm ($\frac{1}{8}$ in) wheel-operated air
release valve. Key-operated air valves will not be allowed
unless specified in the Supplementary Specification.
The Contractor must allow in his tender for taking down
radiators twice and refixing for plasterer and/or painter as
may be necessary during the progress of the work. Backs of
radiators must be 40 mm ($1\frac{1}{2}$ in) from the wall when fixed.
Steel radiators to be provided and fixed on wall brackets.

Invisible panels

31. Provide and fix invisible heating panels of surfaces as
stated and fixed in the positions shown on drawings.
The surfaces of the panels are actual size plus 75 mm (3 in)
on either side. The Contractor may, subject to approval,
modify the size of any panel to suit his required standard but
the total surface in any room must not be reduced. Panels to
be constructed from 15 mm ($\frac{1}{2}$ in) or 20 mm ($\frac{3}{4}$ in) 'Heavy
Weight' steel tubing.
All flow and return connections to panels must be made with
'Heavy Weight' tubing, and G.M. wheel gate valves are to
be fitted to each panel flow connection and lockshield valves
on each return.
Special cover plates and hand wheels with extended spindles
as shown on drawings must be provided and fixed on to the
prepared faces of the chases, for operating the valves. The
faces of the chases will be prepared by the Builder. The actual
position of these controls will be arranged on the site but
will be about 0·60 m (2 ft) to 1·00 m (3 ft) above the floor.
Lockshield valves must be fitted to the bottom of each pair
of risers, together with gland emptying cocks with 15 mm
($\frac{1}{2}$ in) hose connection for emptying-down purposes.

Special care must be taken properly to vent the panel system. At the top of each return riser fit an air bottle 300 mm (12 in) long × 50 mm (2 in) bore surmounted by 6 mm ($\frac{1}{4}$ in) automatic float type air valve, the discharge of which must be carried to a position over the expansion tank.

Expansion tank

32. A properly designed expansion tank of the capacity specified or shown on the drawing must be provided and fixed in the position shown or indicated in the Supplementary Specification.

Every tank shall be fitted with 15 mm ($\frac{1}{2}$ in) approved pattern ball valve and overflow, piped to an approved outside position. No overflow pipe shall be less in size than shown on the drawing or specification.

Supports for tanks will, unless otherwise specified, be provided and fixed by the Builder, except that for contracts where a builder is not employed, or where the Engineering Contractor is the Main Contractor, he will provide and fix such supports.

Connect expansion tank to nearest approved cold water supply feed from the Town main.

Every expansion tank shall be fitted with a 1·6 mm (16 s.w.g.) thick galvanised mild steel cover.

Tanks shall be constructed from mild steel plate, galvanised after manufacture and shall be of the following gauges:

Tank capacity		Plate thickness*	
Litres	*Gal*	*mm*	*B.G.*
45 to 230	10 to 50	1·6	16
275 to 450	60 to 100	2·0	14
560 to 1 125	125 to 250	2·4	12

*Refers to metal thickness before galvanising.

Cold water storage tanks

33. Cold water storage tanks shall be constructed from mild steel plate galvanised after manufacture and all such tanks must be fitted with loose covers of similar material of 1·6 mm thick and stiffened where necessary. Where tanks are more

than 1 m (3 ft) long the cover shall be in two or more overlapping sections each approximately 1 m (3 ft) long. Each
cover or cover section must be fitted with two metal lifting
handles made from 10 mm ($\frac{3}{8}$ in) round iron, galvanised.

Connect tanks to main water service and to supply system as
shown on the drawings and/or indicated in the Supplementary
Specification, if any. Supply and fix overflow pipe or pipes
from tank or tanks to approved outside position. No overflow
pipe shall be less than the size indicated on the drawing or
specification.

Tanks must be made to fulfil the following requirements:

Tank capacity		Plate thickness*		Overflow size	
Litres	Gal	mm	B.G.	mm	in
up to 230	up to 50	1·6	16	32	$1\frac{1}{4}$
275 to 450	60 to 100	2·0	14	40	$1\frac{1}{2}$
560 to 1 125	125 to 250	2·4	12	40	$1\frac{1}{2}$
1 250 to 2 730	275 to 600	3·2	$\frac{1}{8}$	50 to 65	2 to $2\frac{1}{2}$
3 180 to 4 500	700 to 1 000	4·8	$\frac{3}{16}$	80	3

*Refers to metal thickness before galvanising.

Where sectional tanks are required their particulars will be
found in the Supplementary Specification.

Every tank must be supplied with a ball valve of approved
design and size and other tappings as specified or shown on
the drawings. All tanks of 900 litre (200 gal) and over must
be stayed in an approved manner as defined in B.S. 417.

Hot water storage cylinders and indirect cylinders, steam calorifiers

34. New hot water storage cylinders, tanks or other vessels
for direct or indirect hot water supply must be provided and
fixed as indicated in the Supplementary Specification and/or
shown on the drawings. Pressure test certificates from the
manufacturers must be sent by letter to the Engineer before
any pressure vessel of this type is fixed.

Where more than one cylinder or other hot water storage
vessel is to be fixed to serve the same system, stopvalves must
be fitted to enable any vessel to be emptied completely and
isolated from its neighbours, without interrupting the hot
water supply to the building or buildings. These valves must
be fitted whether shown on the drawing or included in the

Supplementary Specification or not, unless their omission is specially called for on the drawing or in the Supplementary Specification.

The Contractor must arrange the cylinder fixing height to ensure the greatest possible circulating head, the cylinders being fixed as high as possible, and the horizontal centre line of the boiler or other heater as low as possible.

The Contractor must supply all fixings, supports or brackets other than brickwork or concrete. Where vessels are supported upon steel or iron cradles or other supports, a soft metal liner must be rigidly fitted between all bearing surfaces of the vessel and the cradle or other support. Actual building-in of supports will be done by the Builder. Where a builder is not employed on the project the fixing must be included by the Contractor as indicated in Clause 32.

All hot water storage vessels must be fitted with tappings or flanged connections for primary flow and return, secondary flow and return, cold feed, altitude gauge and thermometer, emptying cock and in the case of steam-heated calorifiers, a tapping for a safety valve.

Make all pipe connections to the hot water storage vessel or vessels to suit the pipe sizes on the drawing.

Supply and fit to all hot water storage vessels:

One combined altitude gauge and thermometer in 120 mm ($4\frac{3}{4}$ in) diameter stove-enamelled steel case and screwed for 15 mm ($\frac{1}{2}$ in) gas thread.

One 25 mm (1 in) empty gland cock with hose connection.

For steam calorifiers provide and fit spring-loaded relief or safety valve.

Primary steam calorifiers or steam calorifiers of the storage pattern must also be fitted with connections for steam and condensate, in addition to all water connections. The Contractor must also include for connecting the calorifier to the steam, condensate, and water systems, in pipe sizes indicated on the drawings, including all special valves and steam traps, gauges, stopvalves, and steam trap by-pass arrangement for every steam trap.

All cylinders and other cylindrical storage vessels must have bolted heads. Rectangular vessels must have ample manholes.

The equipment shall be the grade given in the Supplementary Specification and shall comply with the appropriate British Standard as follows.

Direct Cylinders. Galvanised mild steel.   B.S. 417
,,   ,,   Copper.   B.S. 699
Indirect Cylinders. Galvanised mild steel.   B.S. 1565
,,   ,,   Copper.   B.S. 1566
Steam Calorifiers for heating.   B.S. 3274
,,   ,,   for hot water supply.   B.S. 853

Boilers

35. New heating boilers and hot water supply boilers of approved manufacture as indicated in the Supplementary Specification and/or shown on the drawings must be provided and fixed.

Before erection all heating and hot water boilers must be given two coats of aluminium paint including all furnace heating surfaces. Where secondhand boilers are supplied to specification, they must be scraped and wire brushed down to the metal and given two coats of aluminium paint as for new boilers.

Every boiler must be fitted with a safety valve, combined altitude gauge and thermometer, as in Clause 34, empty-down cock with hose connection to fit boiler tapping and all necessary cast iron smoke pipe with soot cleaning doors. The Contractor must supply a suitable metal sleeve for building into the flue by the Builder. The space between the sleeve and the boiler smoke pipe must be packed with asbestos rope and finished with fire cement.

Where two or more boilers are installed the Contractor must provide and fix stopvalves to enable one or more boilers to be isolated without interrupting the heating or hot water service and this shall apply whether shown on plans or specified or not. In addition to a safety valve, an open pipe must be taken from all heating and hot water boilers to terminate over the expansion tank or cold water storage tanks, unless other venting arrangements are specified or shown on the drawings.

On single and multiple boiler installations where more than one main circuit is taken from the boiler room, every such main circuit shall be fitted with lockshield valves and this shall apply whether shown on plans or specified or not. Boilers serving single-pipe circuits shall not be valved.

The following safety valve and open pipe sizes must be adhered to.

HEATING BOILERS

Heating surface		Safety valve bore		Open pipe bore	
m²	ft²	mm	in	mm	in
up to 19·51	up to 210	20	$\frac{3}{4}$	32	$1\frac{1}{4}$
19·60 to 26·01	211 to 280	25	1	32	$1\frac{1}{4}$
26·11 to 32·52	281 to 350	32	$1\frac{1}{4}$	40	$1\frac{1}{2}$
32·61 to 37·16	351 to 400	40	$1\frac{1}{2}$	50	2

HOT WATER SUPPLY BOILERS

Heating surface		Safety valve bore		Open pipe bore	
m²	ft²	mm	in	mm	in
up to 13·01	up to 140	20	$\frac{3}{4}$	32	$1\frac{1}{4}$
13·10 to 17·19	141 to 185	25	1	32	$1\frac{1}{4}$
17·28 to 21·37	186 to 230	32	$1\frac{1}{4}$	40	$1\frac{1}{2}$
21·46 to 26·01	231 to 280	40	$1\frac{1}{2}$	50	2

The Contractor must provide for building into chimney base by a builder, a cast iron soot door and frame of the largest possible size permitted by the internal dimensions of the chimney or flue. Where specified in the Supplementary Specification, provide and fix a draught stabiliser.

Flue pipes and flue pipe fittings for gas boilers and gas fired air heaters shall be of asbestos cement, factory treated with vinyl acetate.

Outside flue pipes for gas fired equipment shall also be insulated as laid down in the Supplementary Specification.

Supply and hang on an iron or steel rack on the boiler room wall in approved position a complete set of stoking tools for each boiler or set of boilers of each rating or size. The Contractor shall supply and fix the racks. Each set of tools to comprise:

One shovel, one slicer bar, one poker, one scraper, one round flue brush, one flat flue brush, one pair clinker tongs with revolving claw.

Supply and fit a full set of fire bars to each boiler.

All boilers to be complete with usual fittings, dampers, and mountings whether mentioned in the Supplementary Specification or not. Where required thermostatic damper controls will be included in the Supplementary Specification or the specification for fuel-burning equipment.

Circulators and pumps

36. Provide and fix in positions shown on drawings, or indicated, electrically-driven pumps or circulators of approved make.

The Contractor must submit for approval before ordering, the name, make, size, speed, power and overall efficiency of each machine.

The pump capacity must be as stated in the Supplementary Specification. Where two or more pumps or circulators are to be installed the Contractor must make full arrangements for valving to secure complete isolation and removal of any machine without interruption to the services, whether shown on the plans or not.

Every machine must be designed for super-silent running.

Machines for hot water supply must be designed with a suitable overload characteristic to enable sudden heavy hot water draw-off to take place without damage to pump or motor.

Each machine must be provided with motor and starter, stop-valves, drain valve, counter flanges, lubricators, set of spanners and sufficient packing for one complete repacking of the glands.

The Contractor shall include for fixing the machines and starters and for connecting to the piping system as shown on the drawings or as instructed on the site. Electric wiring will normally be carried out under a separate contract except where indicated otherwise in the Supplementary Specification and/or on the drawing.

Anti-vibration

37. The Contractor shall include for providing and fixing anti-vibration couplings to all pipe connections to and from the pump or pumps.

The Contractor shall provide and fix under all pumps and circulators and all machines, an approved anti-vibration pad or other approved anti-vibration mountings.

Thermal insulation

38. *Section (a)—Steam Boilers.* The exposed outer shell of the boiler must be covered with magnesia and hard setting asbestos, the magnesia to contain not less than 85 per cent hydrated carbonate of magnesia or more than 15 per cent asbestos fibre.

The covering is to be applied in not less than three coats, trowelled to a smooth surface and painted two coats of heat resisting enamel to approved colour. Thickness of covering to be as follows:

Boiler pressure		Magnesia		Hard setting asbestos	
bar	lb/in²	mm	in	mm	in
up to 3·45	up to 50	45	$1\frac{3}{4}$	19	$\frac{3}{4}$
3·52 to 6·90	51 to 100	60	$2\frac{1}{4}$	19	$\frac{3}{4}$
6·96 to 10·34	101 to 150	70	$2\frac{3}{4}$	19	$\frac{3}{4}$

Note: $lb/in^2 \times 6\cdot895 \times 10^{-2} = bar$

In the case of horizontal boilers of the Lancashire, Cornish or Economic type, the front end plate is not to be covered. The covering on all boilers must be finished with canvas and painted as before specified.

When steel or iron flue connections are specified to be covered, special high temperature non-conducting materials 65 mm ($2\frac{1}{2}$ in) thick must be used, reinforced with wire netting and painted two coats approved heat-resisting colour.

Section (b)—Heating and Hot Water Boilers. Hot water heating and hot water service boilers must be covered with magnesia and hard-setting asbestos. The magnesia to contain not less than 85 per cent hydrated carbonate of magnesia or more than 15 per cent asbestos fibre. The covering to be applied is not less than three coats, trowelled to a smooth surface and painted two coats of heat-resisting enamel to approved colour. Thickness of covering to be as follows:

Magnesia 38 mm ($1\frac{1}{2}$ in); hard-setting asbestos 13 mm ($\frac{1}{2}$ in). When steel or iron flue connections or pipes are specified to be covered, special high temperature non-conducting material 65 mm ($2\frac{1}{2}$ in) thick must be used, reinforced with wire netting and painted two coats of approved heat-resisting enamel to approved colour.

All work under this specification must be finished with canvas and painted as before specified.

Section (c)—Cylinders and Calorifiers. Hot water storage cylinders and heating and hot water calorifiers or indirect cylinders must be covered with magnesia and hard-setting asbestos. The magnesia to contain not less than 85 per cent hydrated carbonate of magnesia or more than 15 per cent asbestos fibre.

The covering is to be applied in not less than three coats trowelled to a smooth surface and painted two coats of heat-resisting enamel in approved colours.

Thickness of covering to be as follows:

Magnesia 38 mm (1½ in); hard-setting asbestos 13 mm (½ in).

All work under this section must be finished with canvas and painted as before specified.

Section (d1)—Steam Pipes and Condense Pipes. All steam and condense pipes specified are to be covered with magnesia and hard-setting asbestos. The magnesia to contain not less than 85 per cent hydrated carbonate of magnesia and not more than 15 per cent asbestos fibre. The covering is to be applied in not less than three coats, trowelled to a smooth surface and painted two coats of heat-resisting enamel to approved colour.

Thickness of covering to be as follows:

Pipe size		Magnesia		Hard setting asbestos	
mm	*in*	*mm*	*in*	*mm*	*in*
80 to 150	3 to 6	38	1½	13	½
15 to 65	½ to 2½	32	1¼	13	½

Where covering is likely to become damaged it must be finished with canvas.

Section (d2)—Steam and Condensate Pipes. All steam and condensate pipes must be covered with glass silk sectional 25 mm (1 in) thick with canvas covering and metal bands and finished two coats of heat-resisting enamel to approved colour.

Section (e1)—Heating and Hot Water Pipes. All heating and hot water pipework specified must be covered with magnesia and hard-setting asbestos. The magnesia to contain not less than 85 per cent hydrated carbonate of magnesia and not more than 15 per cent asbestos fibre. The covering to be applied in not less than three coats trowelled to a smooth surface and painted two coats of heat-resisting enamel to approved colour.

Thickness of covering to be as follows.

Pipe size		Magnesia		Hard setting asbestos	
mm	in	mm	in	mm	in
65 to 150	2½ to 6	25	1	13	½
15 to 50	½ to 2	20	¾	13	½

Where covering is likely to become damaged it must be finished with canvas.

Section (e2)—Heating and Hot Water Pipes. All heating and hot water pipes specified must be covered with glass silk sectional 19 mm (¾ in) thick with canvas covering and metal bands, and finished two coats of heat-resisting enamel.

Section (f1)—Cold Water Pipes. Cold water pipes of all sizes where fixed outside must be covered with:

Three layers of 13 mm (½ in) felt with overlap joints. One layer of hard-setting asbestos 25 mm (1 in) thick. One layer of plastic bitumen applied hot, 4 mm ($\frac{3}{16}$ in) thick.

Cold water pipes of all sizes where fixed inside must be covered with:

Two layers of 13 mm (½ in) felt with overlap joints. One layer of hard-setting asbestos 19 mm (¾ in) thick. Two coats of enamel to finish.

Section (f2)—Cold Water Pipes. Cold water pipes must be covered as follows:

50 mm (2 in) diameter and over—50 mm (2 in) cork sectional bitumen sealed and further protected by a 4 mm ($\frac{3}{16}$ in) thickness of bituminous compound.

40 mm (1½ in) diameter and under—38 mm (1½ in) cork sectional bitumen sealed and further protected by a 4 mm ($\frac{3}{16}$ in) thickness of bitumen compound.

Insulation, General. For all water, steam, and condensate lines, the covering must be taken up to the flanged joints, leaving sufficient room for the withdrawal of the bolts without interfering with the adjacent insulation.

When special finishes are required for glass silk insulation, requirements will be indicated in the Supplementary Specification.

Where glass silk insulation is specified for cylinders or boilers, the thickness must not be less than 50 mm (2 in) and the finish must be hard and dustproof and the covering secured with metal bands.

All outside steam and hot water pipes, irrespective of type of covering, must be finished with three layers of bitumen sheeting secured with galvanised bands at intervals of 300 mm (12 in).

Expansion of pipes

39. Pipe lines should normally be laid or hung to allow for expansion (within the low temperature range encountered in heating and hot water work) to take place. The Contractor will be solely responsible for seeing that pipes are installed in the best way to allow freedom of expansive movement, and additional pipe fittings must be installed to assist this where necessary.

When special expansion joints are required they will be called for in the Supplementary Specification and/or shown on the drawings.

Sump pump

40. Where specified or shown on the drawings, the Contractor must provide and fix a hand-operated or electrically-operated pump for pumping water from the boiler room or other basement room to the nearest suitable gulley at ground level.

The Contractor must provide and fix the pump, starter, suction and delivery pipework, foot valve and strainer, priming and draining arrangements and all fixings and fittings necessary to enable the machine to function effectively and efficiently.

Electrical work

41. Electrical work must comply in all respects with the current edition of the Institution of Electrical Engineers *Regulations for the Electrical Equipment of Buildings.*

Gas-operated kitchen and other equipment

42. When gas-operated equipment of any kind is shown on the drawing or indicated in the Supplementary Specification, the Contractor must include for providing, fixing, and connecting to the gas supply from the meter position or other position indicated, and must include also for providing and fixing all pipework, valves, and fittings and making all connections necessary to enable the equipment to be operated effectively and efficiently. To control the supply to every piece of gas-consuming equipment, the Contractor must provide and fix a separate gas governor and a stopcock.

Where solid fuel cookers are specified or shown on the drawing, the Contractor must supply and fix as for gas equipment.

Gas and water service pipe

43. The gas and water service pipes into the building will, unless specified otherwise, be installed by the Gas or Water Undertakings and the cost, unless specified otherwise, will be included as a Provisional Amount in the Main Contract.

Removal of old materials and debris

44. Unless specified otherwise or indicated on the drawings, the Contractor must dismantle and remove all old equipment, pipework, fittings, and materials which will become redundant when the installation as shown and specified is completed. The Contractor must state in his tender the sum he will pay to the Owner for such equipment and materials.

The Contractor shall remove all debris caused by his work as it accumulates and on completion leave all parts clean.

Overtime

45. When submitting his tender the Contractor must allow for all overtime payments to workmen and staff necessary to complete the installation as shown on the drawings and specified. This applies particularly to existing buildings where it becomes necessary to execute certain work without interrupting the routine and services to the occupants. Extra costs will not be allowed for overtime where it can be shown that the Contractor should by his experience have allowed for the cost in his tender. Should the Contractor require guidance on this point when preparing his tender, he should make early communication with the Engineer, who will be glad to advise.

Damage

46. The Contractor will be held to be entirely responsible for damage that may occur to any property and/or plant caused by his workmen or any other representative of his firm, whether by accident or neglect.

Should any such damage occur the Contractor must reinstate the damaged part or parts at his own cost so that on completion all parts shall be satisfactory.

Co-operation

47. The Contractor must co-operate with and attend on other trades as and when required.

Type and quality of materials

48. Where names of manufacturers are mentioned in any specification, or in the drawings or schedules, the Contractor is reminded that this is not an instruction but is meant to indicate only type and quality, and provided these requirements are wholly satisfied, fittings, equipment, etc., made by any approved manufacturer will be accepted, at the discretion of the Engineer, whose written authority must be obtained before the materials or equipment are ordered.

Section B

Ventilation

General clauses

1. General clauses 1 to 10, 12 to 14, and 16, in Section A shall apply.

Fans (propeller)

2. Propeller fans shall be of the streamline pattern, having not less than three blades, properly balanced and mounted on the extended spindle of an electric motor. The fans and motor are to be supported on a cast iron or steel mounting ring or frame by three or more cast iron or steel arms. Propeller fans must be mounted on built-in wooden frames, the fan mounting being secured to the wood frame with not less than three steel bolts with suitable washers, nuts and lock nuts. Corsil rings must be fitted between the fan frame and the wooden frame in all cases. Propeller fan duties to be as stated in the Supplementary Specification, and to ensure quiet running the top speed should not exceed 5·08 metres per second (1000 ft/min).

Weather screens

3. Metal weather screens, hoods or shutters, as specified in the Supplementary Specification shall be provided and fixed to all propeller fans in outside walls. Every propeller fan shall be fitted with an approved wire guard on the inlet. Fan-powered roof extract units shall be as indicated in the Supplementary Specification.

Fume fans

4. Fans for fume cupboards and all positions where corrosive fumes are to be extracted, shall be manufactured throughout from Rigid P.V.C. (polyvinyl chloride) and be resistant to weathering, moist gases, and corrosive fumes at temperatures up to 60°C.

Axial fans

5. Axial or aerofoil type fans shall in all cases be supported externally on anti-vibration mountings as approved by the

Engineer, and to ensure the minimum transmission of fan noise along the ducting system, rot-proofed canvas sleeves and matching angle flanges must be used to connect the fan of the ducting system.

Fan casing

6. The casing of axial fans must extend over the fan and motor, and must be of heavy gauge steel, reinforced for extra strength.

Fan accessibility

7. Where access for inspection and maintenance is unhindered by any machinery or structure, the fan casing shall be fitted with two 90° inspection doors. In positions where maintenance and frequent inspection may be difficult, the fan and motor must be mounted on hinges off the main casing to enable the machine to be swung out for inspection and maintenance.

Electrical terminal box

8. Every axial fan shall be fitted with an external terminal box, mounted on the fan casing to receive the supply cables.

Lubrication

9. External lubricators shall be fitted to every axial fan.

Noise reduction

10. Silencers of approved design and manufacture shall be supplied and fitted by the Contractor, in positions indicated on the drawings.

Automatic damper

11. Where axial fans are operating in parallel, and where stand-by machines are employed, the Contractor shall provide and fix a suitable butterfly damper at the outlet of each fan, designed to close when subjected to back pressure with the fan stationary, and to open fully when the fan is running.

Centrifugal fans

12. Unless specified otherwise in the Supplementary Specification, centrifugal fans for general ventilating purposes shall be single inlet, single width, slow-speed multi-vane machines of approved manufacture.

Fan casing

13. Fan casings shall be built with a steel plate scroll welded to steel plate sides, supported by heavy cast iron side frames, upon which the bearing stool is mounted.
The side frames must be constructed to allow the most convenient angle of discharge.
To facilitate withdrawal of the impeller, the opening in the steel casing to which the cast iron supporting frames are secured must be larger in diameter than the impeller.

Fan drive

14. All centrifugal fans shall be fitted with open vee belt drives, unless varied by the Supplementary Specification.

Fan bearings

15. For general ventilation centrifugal fans shall in all cases be fitted with ring-oiling, self-aligning bearings of ample strength and dimensions.
A large capacity oil well with facilities for replenishing and inspection, and an oil level indicator must be provided.
Efficient oil seals, to prevent oil creep along the fan shaft and to prevent dust entering the bearing ends, must also be fitted.
The use of ball bearings will be allowed in boiler houses, and other positions where noise is not so objectionable.
Unless stated otherwise in the Supplementary Specification, the use of sleeve bearings as specified above shall be implied in all cases.

Fan impeller

16. The fan impeller shall be of the *curved forward* type and shall be statically balanced before it is built into the fan casing, and shall be fitted to the fan shaft with a sunk key of standard dimensions.

Air noise at fan discharge

17. To ensure quiet running the mean air velocity at the smallest area of discharge shall not exceed 9 metres per second (1800 ft/min).

Fan mountings

18. Anti-vibration mountings of approved manufacture must be used in all cases, the design to be approved by the Engineer.

Connection to duct system

19. Rot-proof canvas sleeves must be used for connecting the ductwork to the inlet and discharge sides of the fan.

Electricity supply and control gear

20. Particulars of the electricity supply are included in the Supplementary Specification.
Types, sizes and quality of all switch and fuse gear, including starters, shall be as laid down in the Supplementary Specification and/or shown on the drawings.

Ductwork

21. The installation shall comply with H.C.V.A. Specification DW 121, and the arrangement and layout of ductwork shall be as shown on the drawings. The whole of the ductwork system, including the fans, shall be properly earthed, and where flexible connections are used earth continuity must be maintained by bonding the metal ducts or other equipment. The whole of the ductwork, unless specified otherwise, must be of galvanised steel, of the thicknesses given below before galvanising.
All ductwork for outside fixing shall be not less than 16 gauge, and shall be galvanised after manufacture.

Duct joints

22. Jointing and stiffening of ducts shall be as stated in the Supplementary Specification and where slip joints or angle-iron flanged joints are specified the work shall comply in all respects with the *I.H.V.E. Guide*, and H.V.C.A. Specifications.

Rectangular ducts Length of longer side		Thickness of metal		Circular ducts Diameter		Thickness of metal	
in	*mm*	*mm*	*B.G.*	*in*	*mm*	*mm*	*B.G.*
up to 24	up to 600	0·6	24	up to 20	up to 500	0·6	24
over 24	over 600			over 20 up to 30	over 500 up to 750	0·8	22
and up to 40	and up to 1 000	0·8	22	over 30 up to 50	over 750 up to 1 250	1·0	20
over 40 and up to 90	over 1 000 and up to 2 250	1·0	20	over 50 up to 100	over 1 250 up to 2 500	1·2	18

Compiled from *The Specification for Sheet Metal Ductwork DW/121*, by permission of The Heating And Ventilating Contractors Association. (See list of recommended publications in Appendix A.)

Ducts constructed of materials other than steel or plastics

23. When the Supplementary Specification calls for duct-work to be constructed of materials other than steel or plastic, such work and materials shall comply in all respects with the recommendations contained in the *I.H.V.E. Guide*, and H.V.C.A. Specifications.

Plastic and fibre ducts

24. Where plastic ductwork is specified, the Contractor must produce the manufacturers' certificate guaranteeing its chemical resistance and general properties.
Fibre ducts may also be used, when specified for forming ducts in concrete structures, or as suspended distribution ductwork in approved circumstances. Where suspended fibre ducts are used on warm air installations due allowance must be made on long runs for shrinkage when the system is initially heated up. The manufacturers' written instructions on this point must be carefully observed.
Plastic or fibre ductwork must not be installed unless specified and approved by the Engineer.
Thickness of plastic sheet ducts must comply with the recommendations of the *I.H.V.E. Guide*, and H.V.C.A. specifications.

Air heaters

25. *Electric heaters* of the open coil or sheathed type as specified in the Supplementary Specification, complete with all electrical connections, contactor and safety cut-outs, shall be provided and fixed.

The heaters must be of approved manufacture and fitted with fully protected phase and neutral bus-bars, so arranged that the heater can be switched in sections to facilitate heat control, and also to enable the heater to be connected for use on either a single-phase or three-phase electrical supply.

Full protection must be afforded against excess heating up and damage due to fan or electrical contactor failure, by the provision of an approved hand re-set type thermal cut-out. To ensure that the heater cannot become energised when the fan is stopped, the fan motor starter and heater contactor must be interlocked.

The heaters must be of the type and electrical loading specified.

Gilled Tube Heaters where called for in the Supplementary Specification shall be constructed of special gilled copper tube, welded into steel headers, mounted into flanged steel casings for connecting to the plant or ductwork. Provision must be made in the heater design to allow free movement of the bottom header, enabling the heater tubes to expand and contract freely. The heaters must be constructed to allow easy extension should this become necessary to provide for increased capacity at a later date.

Air filters

26. Air filters may be of the following types as laid down in the Supplementary Specification:

Cotton wool air filters where specified shall comprise a steel plate casing and filter element consisting of a throw-away cotton-wool pad. The number of cells per battery shall be as specified or shown on the drawing, and shall be built into the inlet wall of the ventilating plant chamber, or shall be suitably flanged for fitting to the inlet side of the fan, or into the duct system as indicated on the drawing.

Viscous air filters shall be of the automatic rotary type or the manually operated oblique type, as stated in the Supplementary Specification.

When the oblique filter is specified it shall consist of the number of cells indicated, each cell consisting of a steel casing containing four filtering elements, arranged in staggered formation. The elements must be easily withdrawable for cleaning, re-oiling and general maintenance.

A set of three cleaning and re-oiling tanks, each on a suitable metal stand, shall be provided to facilitate easy cleaning, rinsing and re-oiling. Each tank to be constructed from sheet steel, and provided with a draining rack, drain hole and plug, and lifting handles. Automatic rotary viscous air filters shall be provided and fixed where specified, and shall comprise oil tank, operating gear with electric motor drive, and filter element. The oil capacity and dimensions to be as specified and/or shown on the drawings.

Air washers

27. Air washers shall be of standard type, and the capacity shall be as specified in the Supplementary Specification. Where the washer is not formed into the building structure, it shall consist of a rectangular galvanised steel casing, flanged at each end for connecting to the duct system, and fitted with glazed inspection doors. The water-circulating pump shall be mounted on a suitable concrete base separate from the washer chamber and connected to the inlet and outlet washer pipes with rubber hose connections. The pump to be arranged to discharge at a pressure of 172 kN/m^2 (1·72 bars) into a horizontal spray header. A battery or batteries of spray nozzles as specified, mounted horizontally into vertical spray pipes to form the washer, to be mounted over the water tank which forms the base of the washer. A suitable water strainer tank fitted with loose lid must be installed between the water tank outlet and the pump inlet.

The water tank supply shall be controlled by a 15 mm ($\frac{1}{2}$ in) ball valve, with an additional 25 mm (1 in) valved inlet for quick filling.

On the inlet side of the washer, and immediately before the spray nozzles, a perforated inlet plate must be provided to break up the incoming air for even distribution inside the washer.

Eliminator and scrubber plates of ample capacity, to remove entrained moisture from the air after leaving the washer chamber, must be fitted. Provide and fit a valved sparge pipe for cleaning the scrubber.

Suitable metal walk-ways must be provided inside the chamber to enable all interior parts to be inspected easily, and the interior must be electrically illuminated.

The various parts of the washer shall satisfy the following specification:

Washer casing, galvanised sheet steel, suitably braced and stiffened with tees and angle, galvanised after manufacture.

Pipework. Galvanised wrought iron pipes and fittings throughout. All internal pipes and washer surfaces galvanised after manufacture.

Spray nozzles. Brass, non-clogging type, in two parts to facilitate cleaning.

Sparge pipe. Galvanised wrought iron.

Water tank. Specially coated for protection against corrosion. All-welded heavy sheet steel construction.

Stopvalves. Gate type, made of gunmetal throughout.

Tank overflow and drain. Push fit stand-up drain socket made from aluminium alloy.

Water strainer. Fitted between main washer tank and strainer tank. A perforated brass plate with galvanised iron frame.

Perforated inlet plate. Sheet steel, 19 mm ($\frac{3}{4}$ in) holes. Galvanised after manufacture.

Eliminator and scrubber plates. Made from galvanised sheet steel.

Walk-way grids. Steel, galvanised after manufacture.

Inspection doors. Sliding windows of armour-plate glass with aluminium alloy frames.

Pump. Electrically-driven centrifugal type, with sleeve bearing. Electrical characteristics as Supplementary Specification.

Pump connection. Armoured rubber hose.

Interior illumination. 60 watt aluminium alloy watertight bulkhead fitting.

Pressure gauge on pump discharge. Graduated 0 to 200 kN/m^2 (2 bars) and fitted with fine regulating valve.

The washer installation must be mounted on a concrete, blue brick, or glazed brick base in accordance with the Supplementary Specification.

Where the air washer chamber is to be constructed of normal building materials such as brick or concrete, the structural and finishing details will be included in the Builder's specification and drawings, and will be fitted internally as specified above and indicated in the Supplementary Specification and on the drawings.

Electrical wiring

28. The Ventilation Contractor shall supply and fix all electrical, time, thermostatic, and other operational controls for ventilation plant and equipment.

From a suitable isolating switch fuse supplied and fixed by the Ventilating Contractor, the Electrical Contractor, under a separate contract, shall carry out all wiring to connect the ventilation controls to the electrical system, unless stated otherwise in the Supplementary Specification, or covered by a Provisional Sum in that specification. Wiring to comply with Section E of the Standard Specification.

Section C

Automatic underfeed Bituminous Coal Stokers

General clauses

1. General clauses 1 to 10, 12 to 14, and 16, in Section A shall apply. Installations to comply with B.S. 749, and C.P. 3000.

Clean Air Act

2. The stoker maker must give a certificate or other written assurance that the machine has been designed to burn bituminous coal at the firing rate stated in the maker's catalogue without causing smoke and furthermore must guarantee in writing that the machine satisfies in every way the requirements of the Clean Air Act current at the time of tendering.

Extent of work

3. The Contractor shall include in his tender for the supply, delivery to fixing position in the boiler room, and complete erection and setting to work of an automatic underfeed bituminous coal stoker or stokers, as detailed in the Supplementary Specification and/or shown on the drawings.

Erection shall include placing the machine on a prepared concrete or brick foundation, fitting to prepared front plate of boiler, setting and connecting of retort in the boiler, assembly of retort and all stoker parts, and all refractory brickwork settings necessary to complete the installation.

The concrete or brick foundation slab for the machine will be formed by the Building Contractor under a separate contract, unless stated otherwise in the Supplementary Specification, or provided for by a Provisional Sum in the Supplementary Specification.

Controls

4. The stoker manufacturer shall supply and fit all thermostats, time switches, programme controls, auto-starters, and electrical isolating switch and fuse gear, mentioned in the Supplementary Specification.

Electric wiring from the isolating switch fuse in the boiler room to starter, to motor, and to thermostatic and time controls, will be carried out by the Electrical Contractor under a separate contract, unless specified otherwise or covered by a Provisional Sum in the Supplementary Specification.

Electric wiring

5. Electric wiring for the automatic stoker installation shall be executed by an established electrical contractor. All wiring work is to comply strictly with the terms of Section E of the Standard Specification, except where it is varied in the Supplementary Specification.

Materials and equipment not made by the stoker manufacturer

6. Thermostatic and time controls, and other electrical and mechanical equipment not made by the stoker manufacturer, must be approved in writing by the Engineer.

Motors and starters

7. Motors and control equipment shall be totally enclosed and manufactured in accordance with the current British Standard Specification.

All motors to be fitted with ring-oiled sleeve bearings. All motors shall be continuously rated. Single-phase motors shall be wound for 230/250 volts, 50 hertz, and three-phase machines for 400/440 volts, 50 hertz.

Motor sizes shall not be lower than those specified in the following table, unless approved otherwise by the Engineer in writing:

Hopper stoker rating (heat flow)		Motor rating		Bunker stoker rating (heat flow)		Motor rating	
kW	1 000 Btu/h	kW	hp	kW	1 000 Btu/h	kW	hp
up to 29·31	up to 100	0·37	$\frac{1}{2}$	up to 205	up to 700	0·56	$\frac{3}{4}$
up to 293	up to 1 000	0·56	$\frac{3}{4}$	up to 293	up to 1 000	0·75	1
up to 366	up to 1 250	0·75	1	up to 330	up to 1 125	0·93	$1\frac{1}{4}$
up to 733	up to 2 500	0·93	$1\frac{1}{4}$	up to 733	up to 2 500	1·49	2

Note: Btu/h × 2·931 × 10^{-4} = kW

hp (motor horsepower) × 7·457 × 10^{-1} = kW

Power transmission

8. The gearbox shall be of standard design, and shall be protected by a lightly loaded shearing pin of soft metal.

Regulation of coal feed

9. Regulation of the coal feed shall be obtained mechanically, or electrically, as directed in the Supplementary Specification, and to facilitate release of a jammed coal tube easy manual operation of the feed screw must be possible.

Fume seepage

10. A supply of air under pressure must be fed to the coal tube to prevent fume seepage to the coal hopper.

Hopper capacity

11. The minimum capacity of any coal hopper shall be sufficient to maintain a fuel supply to the furnace at the maximum firing rate of the machine, for a period of not less than 8 hours.
Provide adequate access doors for cleaning out the hopper and its base.
The hopper shall be fitted with a totally sealed lid, unless an open top hopper is called for in the Supplementary Specification.
The stoker rating in Btu/h shall be based on the use of coal having a gross calorific value of 29 000 kJ/kg, and 65 per cent thermal efficiency.

Maximum continuous coal feed

12. From the above it is understood that the maximum hourly continuous coal feed in kg/h, for each stoker shall be equal to:

$$\frac{\text{Boiler rating in kW} \times 3\,600}{29\,000 \times 0 \cdot 65}$$

Fuel size

13. The machine shall be capable of burning coal of moderate coking quality at its full specified rate, and of the following maximum sizes:

Stoker rating (heat flow rate)		Coal size (maximum)	
kW	1 000 Btu/h	cm³	in³
117	up to 400	16	1
352	up to 1 200	20	1¼
733	up to 2 500	25	1½

Note: 1 in³ = 16·387 cm³

Coal feed

14. The coal tube and conveyor screw must be designed to minimise the risk of coal jamming.

Retort

15. The retort shall be of conventional design, and of very robust construction with easily removable, but securely locked, deep web tuyères.

The air inlet ports must be adequately proportioned to ensure the correct air supply to the furnace, and to provide the maximum cooling effect to the grate.

Bunker feed machines

16. Bunker feed stokers where called for in the Supplementary Specification shall include all the general features of the hopper type machines as specified in the preceding clauses, except that a special worm shall be fitted to extend into the fuel bunker.

The drive shall consist of a special heavily designed gearbox mounted on top of the coal tube arranged to drive, by means of a sleeve-protected shaft, a simple secondary gearbox in the coal bunker or alternatively the coal worm shall comprise two separate units, each with its own drive, one to extract the coal from the bunker and convey to a second worm which delivers it to the combustion chamber.

The coal bunker gearbox must be of very heavy construction, capable of running for extended periods without maintenance or lubrication check. Other driving arrangements will be considered by the Engineer, for approval.

Air supply

17. The forced-draught fan shall be of the centrifugal multi-vane type, rated to match the coal feed of the machine based upon the formula given in Clause 12.

The maximum output of the fan in m^3/sec, when operating against the static head of the air delivery duct, the retort, and the burning fuel shall not exceed an amount equal to the maximum hourly coal feed multiplied by 4.

Fan volume control shall be by speed regulation, or by air control at the fan inlet. Damper control on the outlet side of the fan will not be accepted.

The chimney will be designed to provide sufficient draught to deal with the products of combustion at full boiler load, and the fan must deliver air to the furnace, properly to match all rates of coal feed from minimum to maximum, under conditions of balanced pressure in the furnace.

Tests and running instructions

18. The stoker manufacturer shall include in his tender for carrying out complete firing tests (with his own labour) to prove the stoker output and efficiency, under installed conditions, in the presence of the Engineer or his representative.

The stoker manufacturer shall also provide operational instructions to the Owner or his employee. Such instruction to extend over one full working day, and to include fitting new shearing pins, clearing blocked coal tube, removing and cleaning grate tuyères, lubrication, changing rate of coal feed, adjusting air controls, correct furnace conditions, cleaning furnace, adjusting and setting thermostats and time controls, re-kindling, and full information to enable the attendant to obtain the best possible results from the plant.

Types of automatic stoker

19. The foregoing clauses relate to solid fuel heating installations employing sectional boilers and underfeed stokers, which is the most common central plant combination for space heating with solid fuel.

Selection of an automatic stoker for a particular project will be made by the Engineer, and will have regard to the heating load, type of boiler, grade of fuel to be burned, and financial considerations.

If the requirements call for other than underfeed equipment, suitable machines will be chosen from the undermentioned range, and will be detailed in the Supplementary Specification.

(1) Gravity feed stokers using low grade fuels including coke breeze, industrial anthracite grains, and suitable bituminous coals.

(2) Chain and travelling grate stokers burning untreated or washed bituminous coal smalls, and which may be modified to burn coke and coke breeze.

(3) Coking stokers (Low Ram Stokers) applied mainly to large shell type boilers, to burn graded coals and also a wide range of washed small coal.

(4) Sprinkler stokers to burn graded coals, viz. doubles, singles, and peas.

Section D

Fully Automatic Oil Burners

General clauses

1. General clauses 1 to 10, 12 to 14, and 16, in Section A shall apply. Installations to comply with C.P. 3002 and B.S. 799.

Extent of work

2. The Contractor shall include in his tender for the supply, delivery to fixing position in the boiler room, and complete erection and setting to work, of a fully automatic oil burner or burners, as detailed in the Supplementary Specification and/or shown on the drawings.

Erection shall include placing the machine on a prepared concrete or brick foundation, fitting to prepared front plate of boiler, assembly in boiler furnace, and all refractory brickwork necessary to complete the installation.

The concrete or brick foundation slab for the machine will be formed by the Building Contractor under a separate contract, unless stated otherwise in the Supplementary Specification, or provided for by a Provisional Sum in the Supplementary Specification. Provide and fix oil storage tank or tanks, and all pipework.

Types of oil burning equipment

3. Oil burners will be chosen from the entire range of available equipment, and depending upon the requirements of the particular scheme, will be detailed in the Supplementary Specification.

For general space and water heating, equipment of the types mentioned below will be installed:

Approximate heat load kW	Type of burner	Fuel
Up to 24	Fully automatic vaporising pot or wall flame burner	28 seconds vaporising oil
25 to 73	Automatic pressure jet burner with on/off control	40 seconds gas oil, or 200 seconds fuel oil
74 to 117	Automatic pressure jet burner with high/low/off control	40 seconds gas oil, or 200 seconds fuel oil
132 to 586	Automatic pressure jet burner with modulating/off control	40 seconds gas oil, or 200 seconds fuel oil

The use of other types of oil burner, notably those in the range of Blast Atomisers, will depend upon load and economic factors, and where their use is considered advantageous, or essential, will be called for in the Supplementary Specification.

Motors

4. Motors and control equipment shall be totally enclosed, and manufactured to comply with the current British Standard Specification.

All motors shall be fitted with ring-oiled sleeve bearings.

All motors shall be continuously rated.

Single-phase machines shall be wound for 230/250 volts, 50 hertz, and three-phase machines for 400/440 volts, 50 hertz.

40 seconds oil				220 seconds oil			
Maximum boiler rating		Motor rating		Maximum boiler rating		Motor rating	
kW	1 000 Btu/h	kW	hp	kW	1 000 Btu/h	kW	hp
29·31	100	0·25	$\frac{1}{3}$	293	1 000	0·37	$\frac{1}{2}$
293	1 000	0·37	$\frac{1}{2}$	586	2 000	0·56	$\frac{3}{4}$

Note: Boiler ratings are based on 65 per cent efficiency with oil at:
  45 500 kJ/kg gross calorific value, 0·84 specific gravity, for 40 sec oil.
  43 700 kJ/kg gross calorific value, 0·93 specific gravity, for 220 sec oil.
  Btu/lb × 2·326 = kJ/kg.

Motor sizes shall not be less than specified in the above table, unless approved otherwise by the Engineer, in writing.

Pump

5. The pump must be designed to deliver oil at the required maximum pressure, and shall be factory-tested and adjusted for the specified duty, and fitted with all pressure-regulating and cut-off valves. Provide and fit a pressure gauge.

Fan

6. The air fan shall be the directly-driven multivane type, designed to deliver the correct volume of air at the right pressure for the burner duty. Adjustable air control shall be provided.

Mounting

7. The complete machine shall be mounted on a mild steel tray supported on levelling screws.

Oil filter

8. On the suction side of the pump, provide and fit a suitable self-cleaning oil filter.

Stopvalves

9. Provide and fix all necessary stopvalves to facilitate maintenance.

Burner oil heater

10. Provide and fit an electric oil heater suitable for standard single-phase wiring, and thermostatically controlled. Provide and fix an oil temperature thermometer.

Ignition

11. The burner head to be of heavy robust construction, with heavy section electrodes fitted in an easily accessible position for quick and accurate adjustment.
Ignition to be by high tension spark, from a continuously-rated ignition transformer.

Controls

12. Provide and fix an approved control box and relay; adjustable main On-Off control thermostat, boiler limit thermostat, flame failure safety device and time switches.
Time switches and manual change-over switches for operational programming shall be laid down in the Supplementary Specification, and/or shown on the drawings.

Burner wiring

13. For each machine a suitable isolating switch fuse will be provided and fixed in the boiler room by the Electrical Contractor who will also wire up to each switch fuse unit.
The oil burner manufacturer will be responsible for all

electric wiring between the isolators referred to above and the burner control box and relay, and from control box to the machine, including all thermostat and flame stat wiring, and must include the cost of this work when quoting for the machine.

If the manufacturer does not employ skilled labour for this purpose, he must include a Provisional Amount to cover the cost of the wiring and must supervise the work, for which he will be entirely responsible. The Engineer will not accept that the General Heating Contractor be held initially responsible for any part of the oil-burning installation and controls. The manufacturer must submit wiring diagrams of the complete installation for the Engineer's approval, and will be responsible for the completed installation.

Materials and equipment not made by the burner manufacturer

14. Control boxes, relays, thermostats, time controls, and other electrical or mechanical equipment not made by the burner manufacturer must be approved by the Engineer, in writing. Samples must be provided for examination if required by the Engineer.

Oil storage tanks

15. Provide and fix on prepared brickwork or concrete supports a steel oil storage tank or tanks, of the capacity laid down in the Supplementary Specification, and in the position shown on the drawing.

The tank shall be rectangular, or cylindrical, with dished ends, or cylindrical with flat ends, as specified.

Each rectangular tank shall be fitted with a 450 mm (18 in) diameter manhole and setpinned cover with metallic asbestos joint ring, and shall be strongly stayed with heavy welded steel stays, and shall be fitted with 50 mm × 65 mm ($2 × 2\frac{1}{2}$ in) and 50 mm × 40 mm ($2 × 1\frac{1}{2}$ in) diameter screwed bosses for filler, vent, sludge, and outlet pipes respectively.

Cylindrical tanks shall be fitted with a 450 mm (18 in) diameter raised manhole, with bolted cover and metallic asbestos joint ring and fitted with 50 mm × 65 mm and 50 mm × 40 mm diameter screwed bosses for filler, vent, sludge, and outlet pipes respectively.

All tanks shall be steel of welded construction, normally designed for a pressure of 35 kN/m² (0·35 bar).

Where tanks are to be fitted in basements where they could become subject to higher pressures than the normal pressure of 0·35 bar, they should be designed to withstand 1·5 times the actual working pressure.

Thickness of plate for normal pressure shall be as follows:

Approx. capacity		Plate thickness (minimum)		
Litres	Gal	mm	Gauge	in
455	100	1·6	16G	
790 to 1 360	175 to 300	2·0	14G	
2 270 to 4 550	500 to 1 000	4·8		$\frac{3}{16}$
9 090 to 45 460	2 000 to 10 000	6·4		$\frac{1}{4}$

Cylindrical tanks shall be mounted upon iron cradles, resting on the brick or concrete supports, and a soft metal liner shall be fitted on the bearing surface between tank and cradle.

At the bearing points of rectangular tanks, a soft metal liner shall be fitted between tank and support.

Tanks shall be insulated as detailed in the Supplementary Specification.

Set horizontal cylindrical, and rectangular tanks with a slope of 40 mm per metre of length, falling towards the drainage point.

Filling pipe

16. Provide and fix an oil filling pipe of size indicated on the drawing, to terminate in the open air in a convenient position for connecting to the delivery tanker's hose. Each tank must be fitted with an independent filling pipe. Provide and fix a stopvalve, drain cock and cap and chain to each tank filling pipe.

Vent pipe

17. To each tank provide and fix an open vent pipe of the same size as the filler pipe. The vent should terminate with a double bend looking down, and fitted with a strong open mesh wire cage.

Drain or sludge valve

18. A gunmetal drain or sludge valve must be fitted to the underside of every tank, at the opposite end of the tank to the outlet pipe.

Suction pipe

19. Provide and fix a valved suction pipe or ring main from the oil storage tank to the burner, in the manner and of the size shown on the drawing.

Heaters

20. Provide and fix in each tank an electric, hot water or steam heater, as specified and/or shown on the drawing.
The heater must be thermostatically controlled, to maintain the oil temperature at 7°C.
Tracer heaters in the form of hot water or steam pipes, or electric heating cable, shall be provided and fixed to oil suction pipes or ring mains where specified.

Fire valve

21. Provide and fix in the suction pipe line near the tank, a glandless type, spring-loaded, or weight-operated fire valve, kept open by the tension of a wire in which a fusible link designed to fuse at a temperature of 68°C is fixed over each boiler combustion chamber. The wire connecting the fire valve and fusible links shall be stranded cable run on suitable pulleys.

Electric wiring
22. All electric wiring associated with the oil burner installation shall comply strictly with the terms of Section E of the Standard Specification.

Tests and running instruction

23. The oil burner manufacturer shall include in his tender for carrying out complete firing tests (with his own labour), to prove the burner efficiency and output under installed conditions, in the presence of the Engineer or his representative.

The burner manufacturer shall also provide operational instruction to the Owner or his employee. Such instruction to include lubrication, stripping of the burner nozzle and ignition electrodes, changing nozzle and electrodes, clearing of day-to-day faults on the machine, setting of thermostats and time controls; information on oil pressure and temperature, flame quality, tank oil level guage reading, sludging of oil storage tanks, clearing oil pipe blockage, operation of fusible fire safety valve, operation of chimney draught stabiliser, and full information to enable the attendant to operate the plant and to diagnose the simple recurring faults to which all such installations are prone.

The period of instruction shall extend over two full working days of 8 hours each.

Section E

Electrical Work Associated with the Heating and Ventilating Installation

General clauses

1. General Clauses 1 to 10, 12 to 14, and 16, in Section A shall apply.

Regulations

2. All electric wiring and apparatus shall satisfy the requirements of the appropriate British Standard Specification in so far as it is not varied by this Specification.

Electric wiring and apparatus is to comply in all respects with the recommendations prescribed by the Institution of Electrical Engineers, Board of Trade, Home Office, Principal Fire Insurance Companies, or any other Authority having jurisdiction.

All wiring and each item of equipment shall comply in all respects with the current edition of the *Regulations for the Electrical Equipment of Buildings* as issued by the Institution of Electrical Engineers, and with the appropriate regulations as issued by the Electricity Commissioners.

Local bye-laws

3. The Electrical Contractor shall ascertain and act in accordance with any Local Bye-Laws and/or Regulations of any other Authority having jurisdiction that may affect the work or operational conditions of any electrical equipment installed under this contract. Any cost incurred in the provision of suppressors, for electrical apparatus requiring same, or testing, shall be covered by the Contractor in his tender.

Materials and workmanship

4. All materials and workmanship throughout the entire contract are to be the best of their respective kinds and to the entire satisfaction of the Engineer or his representative under whose supervision the whole of the work will be carried out, and who will have the power to condemn all faulty materials and workmanship. All rejected materials must be immediately removed from the site and all faulty workmanship made good by the Contractor at his own expense.

Final plans

5. At the conclusion of the scheme a set of drawings will be supplied to the Contractor, upon which he shall show the position of all distribution fuseboards, switches, etc., as actually installed.

Earthing

6. The whole of the conduit system, metallic sheathings, casings to switch and fuse gear and appliances connected with the electrical installation must be earthed strictly in accordance with the latest recommendations of the I.E.E. Regulations for the Electrical Equipment of Buildings, Section D, 14th edition (metric).

Earthing Leads. Earthing conductors shall be in accordance with Table D.2M, page 79 of the 14th edition (Metric 1970) of the I.E.E. Regulations.

Note: Where earthing leads are exposed to the weather phosphor bronze shall be used in preference to copper.

Every connection of an earthing lead to a pipe, conduit, cable sheath, armouring or earth electrode shall consist of a substantial clamp, constructed of either a non-rusting metal such as copper or, where exposed to water or corrosion, of phosphor bronze, and the contact surfaces shall be clean. Where the armouring of a cable is used as an earth electrode or as an earth continuity conductor, it shall be bonded to the lead sheath (if any) of the cable, and the clamp shall be so designed and fixed as to grip the armouring firmly and permanently without damage to the insulation or the lead sheath, but on an armoured lead-covered cable the principal contact of the clamp shall be with the lead sheath.

Note: The armouring of cables will not be relied upon for the purpose of earthing, but shall be bonded to the earthing network where required.

Resistance of earth continuity conductor. The resistance between the earth electrode and any other earthed position on a completed installation, excluding the resistance of the earth leakage circuit breaker if fitted, shall not exceed 1 ohm.

The Electrical Contractor shall include for the supply, fixing, and connecting up of all earth wires, earthing strips, and earth plates and/or earth leakage trips as required for the whole installation, including the main panel and all equipment, etc., to comply with I.E.E. Regulations, Section D, 14th edition (Metric).

Switchgear, fusegear, and control gear

7. *Generally.* All control gear shall be metal clad with ample clearance for wiring up to full capacity in any circumstances and where necessary flanges suitable for direct bolting of sealing boxes and/or glands where paper-insulated or other lead-covered and/or armoured cables are used.

The cases shall have substantial fixing lugs and solid doors. They shall be protected against corrosion in an approved manner.

Bus-bars shall be of ample section for mechanical strength and contact area. The current density shall not exceed the appropriate B.S. rating.

All bolted or clamped connections shall have spring locking washers, and the cable ends shall be sweated into suitable cable sockets or terminals. Terminals bored out to receive cable ends and fitted with set-screws, shall have at least two set-screws gripping each cable.

Dividing fillets shall be of non-warping fire-proof material.

All switch and fusegear shall comply with the current British Standard Specification.

Fusegear. Fuses shall be of the Handshield (H.O.) pattern with all live points effectively shielded.

Fuseboards and sub-distribution boards shall be of sufficient size to accommodate the total number of circuits required in respect of both lighting and power installations and shall be provided with at least two spare fuse ways.

Where neutral, middle wire, or common return of the source of electricity supply is declared 'earthed' by the Electricity authority, single pole and neutral, or triple pole and neutral distribution fuseboards, switch fuses, isolators, socket outlets, etc., as required shall be installed.

Labels. Every switch fuse, switch fuseboard, or set of cutouts shall be fitted with an appropriate permanent indelible label giving either (*a*) the identification letter and number corresponding to a wiring diagram or (*b*) details of the service controlled. In addition every fuseboard shall have a printed and varnished circuit list fixed inside the door.

Control, distribution, and excess current protection gear must be selected and installed to comply in all respects with Section A, of the 14th edition (Metric 1970) I.E.E. Regulations.

Cables

8. Cables should be selected from the appropriate range given in Section B, 14th edition (Metric 1970) Edition of the I.E.E. Regulations.

Wiring

9. Wiring throughout the electrical installation shall be carried out in a workmanlike manner. Outside fittings and current using appliances shall be 'weatherproof' type, so constructed that when installed, rain, snow, and splashings are excluded. All wiring shall be run in heavy steel conduit. All electrical connections to items of equipment which are rigidly fixed shall be connected to the control gear by means of fixed conduit and a short length of flexible metallic tube, and separate earth wires shall be provided.

Conduits and cables shall be fixed in positions where they will not be exposed to rain, dripping water or condensation or accumulations of water or oil, or to high temperatures from boilers, steam pipes or other sources of heat, unless the cables and accessories are adequately shielded or are specially designed to withstand the effects of exposure to water, oil or heat.

All cable ends connected to distribution fuseboards, switches, etc., shall have the insulation taping and braiding carefully cut back and neatly bound up in rubber sleeving.

Conduit and conduit fittings

10. Steel conduits, bends, and couplers shall be strictly in accordance with B.S. 4568: Part 1: 1970 (metric units) and galvanised.

Fittings and components shall be steel, galvanised and shall comply in all respects with B.S. 4568: Part 2: 1970 (metric units).

All boxes shall be fitted with heavy pattern lids of the same material as the box, secured by brass screws. No elbows, inspection or solid shall be used. No inspection bends shall be used.

Except as specifically instructed, all tube work shall be fixed and complete before any conductors are drawn in.

Spout entry boxes shall be used wherever possible, but where not practicable as in certain adaptable and fuse boxes, the conduit connection shall consist of (1) on the inside of the box a hexagon male smooth bore bush, screwed through a clearance hole in the side of the box into (2) on the outside of the box a standard coupler on the end of the conduit. Both bush and coupler shall make a proper metallic joint with the box and where any doubt on this matter occurs, a suitable lead washer or coned brass ring shall be provided on the bush and on the coupler. A suitable lead washer shall be used for all watertight joints.

All screwed joints to have at least six full turns of thread engaged. 'Long-screw' connectors shall be fitted with heavy pattern ring back-nuts.

'Watertight' conduit installations shall have machine-faced joints on all boxes and all fixing holes shall be 'blind' to the interior of the conduit system.

Connections to motors or other apparatus subject to variation shall be made with a length of not less than 100 mm (4 in) of heavy watertight metallic tube sweated into suitable specially screwed conduit adaptors. The metal frame or other parts of apparatus so connected shall be earthed to the conduit system independently of the flexible tube, to the satisfaction of the Engineer.

Drainage holes or tees shall be provided where called for by the Engineer, and all conduit laid to falls to the points as specified on site.

Special attention shall be given to cleaning the threads free from dirt, grease and other foreign matter.

All damage to galvanising shall be properly repaired, and all exposed threads shall be painted after erection. All screwed ends of conduit to have a coat of aluminium spirit paint before and after assembly, to protect the threads from rusting.

All conduit not cast in concrete shall be fixed at not more than 1·0 m centres for 19 mm ($\frac{3}{4}$ in) conduits and 1·22 m centres for larger conduits by suitable screws not less than 25 mm (1 in) No. 8 or equivalent. For surface runs, all galvanised conduit shall be run on distance saddles. For multiple runs of conduit suitable multiple saddles shall be employed. Crampets, nails, and similar driven fixings will not be permitted in any circumstances. All conduit adaptable and fittings boxes shall be fixed by screws independently of the runs of conduit. On 'watertight' installations the box fixing screws shall be external to the box.

Special care

11. Special care shall be taken to prevent water, dirt or rubbish getting into the conduit work during erection. Screwed metal caps or plugs only shall be used for protecting open ends; plugs of wood, waste, paper, etc., shall in no circumstances be used.

On concrete and brickwork

12. On concrete and brickwork all conduit boxes shall be fixed by means of 32 mm ($1\frac{1}{4}$ in) No. 8 screws securely screwed into 'Rawlplugs'. The holes in the concrete and/or brickwork shall be neatly drilled by means of a 'Rawlplug' Rapper.

On steel girders

13. On steel girders all conduit boxes shall be fixed by means of 19 mm × 8 mm ($\frac{3}{4} \times \frac{15}{16}$ in) diameter countersunk headed screws, the girder being drilled and tapped to suit.

On lath and plaster ceilings and partitions

14. On lath and plaster ceilings or partitions the main wood bearers in the case of ceilings and the vertical studding in case of partitions shall be bridged across by a wooden trimmer securely fixed, on to which the conduit boxes shall be screwed by means of 32 mm ($1\frac{1}{4}$ in) No. 8 wood screws.

Note: For each of the above cases the conduit boxes shall be drilled and *countersunk* to take the respective sizes of screws specified.

Conduit details

15. Conduit and *conduit fittings* shall be heavy gauge, screwed, welded, galvanised steel tubes to B.S. specification. All conduits to be of approved material and manufacture throughout.

The use of pressed steel conduit fittings is forbidden except where specifically stated.

Conduit work

16. All conduits chased in walls shall be so recessed that all tubes are finally covered with at least a full thickness of plaster.

The ends of all conduits shall be reamered out and left smooth, and shall be securely fixed into the respective fuseboards, switch boxes, etc., in an approved manner.

The conduit runs may for convenience be run in the heating ducts where possible, but in such cases the conduits shall be arranged as far as possible from heating pipes.

The wiring shall be carried into all electrical equipment and machinery, using a length of fixed conduit and a short length of flexible metallic tube.

Tests

17. The testing of the installation shall comply in all respects with Section E of the 14th (Metric 1970) edition of the I.E.E. Regulations and shall be carried out wholly or in part at the discretion of the Engineer or his representative.

In the case of mineral-insulated metal-sheathed cables being used, such cables shall be given an electrical test before sealing or being covered in the screeding of floors, etc. Such test shall include (*a*) testing between conductors, (*b*) between each conductor and outer sheathing. A 500 volt Megger shall be used for these tests, and the results obtained shall not be less favourable than those specified by the I.E.E. Regulations, i.e. Section E, 14th edition, and the current edition of the Regulations issued by the Electricity Commissioners where applicable.

Tests as prescribed by the I.E.E. Regulations, current edition, must be carried out by the Contractor in the presence of the Engineer. The tests shall include:

  (i) Insulation resistance.
 (ii) Polarity of single pole switches.
(iii) Earth continuity path.
(iv) Effectiveness of earth.
 (v) Full load test which shall be sustained for at least one hour to ensure that all circuits are fused correctly.

The results of all tests shall be reported in writing by the Contractor to the Engineer.

Stoker motors, pumps, and oil burner motors

18. Where stoker or oil burner motors are required, the wiring shall be carried direct into a 15 A T.P. & N. isolator to control each motor installed, and shall terminate therein. The position of each isolator required shall be adjacent to the stoker motor to be controlled.

Where circulating pump motors are required the wiring shall be carried direct into the motor itself and shall be controlled in each case by a 15 A T.P. & N. isolator which will be supplied and fixed by the Electrical Contractor in a position adjacent to the motor to be controlled.

The starters will be supplied under separate contract, but the Electrical Contractor shall include for all fixing and wiring necessary to leave each pump motor in perfect working order. The actual positions of each stoker, oil burner, and pump motor will be determined either by the Heating Contractor or the Engineer.

Unless otherwise specified, wiring for stokers, burners, and circulating pumps shall in all cases be suitable for three phase four wire distribution, and capable of carrying the current consumption of three horsepower per three phase motor.

Sump draining pump motor

19. The wiring shall be carried direct into the sump draining pump motor, which shall be controlled by a single or double pole neon indicator unit to suit the nature of the supply, and supplied and fixed under this Contract.

The motor itself will be supplied under the Heating Contract but the Electrical Contractor shall include for connections thereto, and everything necessary to leave the motor in perfect working order.

Fusing

20. Fuses shall be of such capacity as will allow the full load of each circuit to be carried without undue temperature rise. Fuses for all switch fusegear and distribution fuse boards shall be included for under this contract.

Control, distribution, and excess current protection shall

comply with Section A of the I.E.E. Regulations, 14th (Metric 1970) edition.
Sizes of wire for rewirable fuse gear are given in Table A.1M, Regulation A12.

Identification of cables

21. To permit the easy identification of every single-core, twin-core, and multi-core cable, Regulations B.54 to B.59 inclusive, Section B of the I.E.E. Regulations, 14th (Metric 1970) edition, must be adhered to.

Wiring of fractional horsepower motors

22. Where a number of fractional horsepower motors is grouped on a common final subcircuit the groupings should be arranged so that the aggregate full-load current does not exceed 15 A, and where the supply system is three phase arrangements should be made to protect each motor against the effect of single phasing.

Protection against earth leakage currents

23. Every conductor and item of apparatus shall be effectively prevented from giving rise to the danger of earth leakage currents, by strict application of the conditions laid down in Section D of the I.E.E. Regulations, 14th (Metric 1970) edition.

Flameproof equipment

24. All equipment installed in hospital or sanatoria operating rooms, garages or other situations, specified or unspecified, where inflammable gases are used, must be fully flameproof. The apparatus must be certified 'Buxton Tested' and must comply with the appropriate British Standard Specification.

Motors and starters

25. Motors and starters to be of approved manufacture and to conform to current British Standard Specifications. All wiring to motors from the fuseboard to starters and the inter-connecting wiring from the starters to motors to be p.v.c. in heavy gauge galvanised conduit and all coupled up complete. The positions of motors to be as shown on the drawings, together with the circuit sizes indicated.

Isolating switches

26. Isolating switches shall be galvanised ironclad and shall be provided for every motor and shall be D.P. or T.P. as required.

*   *   *

The Advantages of the Standard Specification

Large Local Authorities, Government Departments, and commercial concerns who are continuously engaged on extensive development, have their own standard specification covering both engineering and building work.

The specification will be similar to that set out in the foregoing pages, and divided into sections covering all the broad aspects of the mechanical and electrical equipment of their buildings, which are common to all building projects.

The specification sections are usually printed separately and issued to the contractors when tenders are invited.

A large organisation requiring tenders for an oil-fired hot water heating installation would send to the tenderers copies of their Standard Specification, Sections A, D, and E, together with scheme drawings and a Supplementary Specification giving such details as boiler sizes, radiator surfaces and patterns, tank sizes, etc., peculiar to the particular job.

An invitation to tender for ventilation work would be accompanied by Sections B and E of the Standard Specification, plus scheme drawings and a Supplementary Specification giving fan sizes and capacities, duct sizes, air grille sizes and types, and any other special instructions which may be necessary for the successful completion of the work.

The Standard Specification is helpful to Owner and Contractor alike, in that it lays down a minimum basic standard of quality acceptable to the Owner, and leaves the Contractor in no doubts as to the general quality of materials and workmanship for which he must allow in his tender, and which will be expected from him should his tender be accepted.

Time spent by technical staff and typists in preparing specifications is also greatly reduced by the proper use of a standard form of specification. This is, of course, an important factor in a busy design office.

SUPPLEMENTARY SPECIFICATION

The following is a typical example of a supplementary specification:

SUPPLEMENTARY SPECIFICATION for a Low Pressure Accelerated Hot Water Heating Installation, to be carried out at the general offices of Messrs. Blank & Co.

Extent of contract

1. The Contractor shall include in his tender for providing and fixing a complete low pressure hot water accelerated heating system as specified hereunder, and as specified in Sections A and E of the Standard Specification and drawings, copies of which have been sent with this specification.

Standard specification

2. This specification must be read in conjunction with the Standard Specification.

Heating boiler

3. Provide and fix in the position shown on the drawing one cast iron sectional heating boiler of approved manufacture, rated at 45·78 kW, and having a boiler heating surface of 3·30 m² (35·5 ft²).
The boiler to have 50 mm (2 in) flow and return connections as indicated on the drawing.

Boiler mountings

4. Provide and fix a full set of boiler mountings and open vent pipe to comply with the Standard Specification, Section A. Empty cocks to be 20 mm ($\frac{3}{4}$ in).

Smoke pipe

5. Provide and fix 200 mm (8 in) diameter smoke pipe and fittings to comply with the Standard Specification. Connection to brick chimney as specified.

Cleaning tools

6. Provide and fix in approved position in the boiler room a tool rack and full set of tools, in accordance with the Standard Specification.

Radiators

7. Provide and fix cast iron radiators of approved manufacture. The radiators to be hospital pattern of the surfaces, sizes, and connections indicated in the radiator schedule on the drawing.

Feed and expansion tank

8. Provide and fix at low level in the roof space, in the position shown on the drawing, a feed and expansion tank, designed to allow for 45 litres (10 gallons) expansion.

Feed and expansion pipe

9. Provide and fix a feed and expansion pipe from the bottom of the tank to the boiler room, and connect to the return main as shown on the drawing. Provide and fix a stopvalve near the tank in the roof space, if shown on the drawing.

Circulating pump

10. Provide and fix in the flow main near the boiler, as shown on the drawing, a pipeline fullway centrifugal pump of approved manufacture. The pump to have a gunmetal body and impeller, and 50 mm (2 in) flanged connections, drilled ready for erection and fitted with a pair of 50 mm (2 in) counter-flanges. The pump must be designed to circulate 0·91 litres (0·2 gallons) of water per second against a total head of 0·09 bar (3 ft).
The pump must be mounted on a set of cantilever wall brackets, and anti-vibration pads must be inserted between the pump and the wall brackets.
Provide and fix isolating stopvalves and by-pass with a lightweight 50 mm (2 in) back pressure valve, as shown on the drawing.

Pump motor and starter

11. The pump motor and starter must be suitable for 415 volts, 3 phase, 50 hertz, a.c. supply, and the electric wiring must be carried out by an approved electrical contractor to comply with Section E of the Standard Specification.

The Heating Contractor shall include a Provisional Sum of £35·00 in his tender to cover the cost of electric wiring to the pump.

Insulation

12. The boiler and all pipework in the boiler room, except the feed and expansion pipe and the open vent pipe, must be thermally insulated with plastic covering to comply with Clause 38 of the Standard Specification.

RADIATOR SCHEDULE

The radiator schedule referred to in Clause 7 of the Supplementary Specification can be shown on the plans (a separate schedule for each floor) or may be embodied in the Specification.

Most Authorities prefer its inclusion with the drawings as in this form it is handy for the estimator, and certainly more convenient for the fitter, who can more readily identify each radiator or other form of heater with its position on the plan.

RADIATOR SCHEDULE, GROUND FLOOR

Rad. No.	Type	Height mm	Width mm	Length mm	Surface m²	Connections mm	Fixing	Remarks
1	Steel (double pane!)	590	67	1 920	5·09	20 T.B.O.E.	Wall brackets	Radiator shelf
2	,,	590	67	1 920	5·09	20 B.O.E.	,,	,,
3	,,	590	67	2 240	5·94	20 T.B.S.E.	,,	,,
4	,,	590	67	1 920	5·09	20 T.B.O.E.	,,	,,
5	,,	590	67	1 920	5·09	,,	,,	–
6	Steel (single panel)	590	29	1 920	2·54	15 T.B.O.E.	,,	–
7	,,	590	29	1 920	2·54	,,	,,	–
8	,,	590	29	1 920	2·54	,,	,,	–

A typical radiator schedule is given here.

This completes the information which is normally furnished by the Owner, through his Architect or Consulting Engineer, to enable the Heating Contractor to prepare a tender.

Standard and Supplementary Specifications and drawings are tender documents, and will later become part of the Contract Documents, to be legally completed by both the Owner and the Heating Contractor, should the latter's tender be accepted.

TERMS AND CONDITIONS OF THE CONTRACT

In addition to the purely technical documents which have been discussed, the invitation to tender will carry further documents dealing with the terms and conditions of the tender, specific instructions governing the progress of the contract, or sub-contract, terms of payment, and all matters affecting the contract relations between Owner and Heating Contractor from date of order to completion, and final acceptance and payment for the work.

If the work is let as a sub-contract, the Heating Contractor will be required to enter into a sub-contract with the Main Contractor in accordance with the Terms and Conditions of the Main Contract relative to Sub-contractors.

Contract Conditions affecting the Tender Price

Certain contract conditions have a direct bearing upon the tender price, and the cost of complying with such conditions will therefore be included in the estimate. Briefly stated, these conditions are:

(1) Where the heating is a sub-contract, the Heating Contractor will, as a Nominated Sub-Contractor, include in his tender an amount equal to $2\frac{1}{2}$ per cent of the quoted price, to be allowed to the Main Contractor as cash discount.

(2) Storage of plant and materials, and protection of work must be covered.

(3) Bond. The contractor may be required to provide at his own expense a bond as security for the due performance of the contract in an amount equal to 5 per cent of the Contract Sum.

(4) The payment of purchase tax where applicable, on articles or materials supplied for the work is the responsibility of the Heating Contractor.

The Variation Schedule

Tenders for heating, hot water supply, ventilation, and similar engineering works are invariably submitted as a lump sum, based upon working drawings and detailed specifications supplied by the Owner, who does not usually furnish a Bill of Quantities for pricing by the engineering contractor.

In such cases the General Conditions of the Contract require completion by the contractor of a Variation Schedule, which comprises a comprehensive list of fittings and materials to be used for the proposed work. The contractor prices and completes the schedule at the time of tendering, or following acceptance of the tender, as may be agreed, and the cost of any subsequent additions or deductions is based on these figures.

Alternatively the contractor may, if his tender is successful, submit a priced Bill of Quantities based upon the drawings and specifications, which of course must be adequate for this purpose, and upon which contract variations can be priced in accordance with Section XVI, Standard Method of Measurement.

A typical variation schedule in common use is given below.

SCHEDULE RATES FOR ADDITIONS AND DEDUCTIONS FOR WORK CARRIED OUT AS FOR CONTRACT

A. Price per metre run for supplying and fixing mild steel tubing (inclusive of sockets and brackets). (a) Galvanised; (b) Medium quality; (c) Heavy quality.

Size		Price			Size		Price		
mm	*in*	*(a)*	*(b)*	*(c)*	*mm*	*in*	*(a)*	*(b)*	*(c)*
150	6				40	$1\frac{1}{2}$			
125	5				32	$1\frac{1}{4}$			
100	4				25	1			
80	3				20	$\frac{3}{4}$			
65	$2\frac{1}{2}$				15	$\frac{1}{2}$			
50	2				10	$\frac{3}{8}$			

B. Price each for pipe fittings and valves (as specified), supplied and fixed.

Tees

Size		Price	Size		Price
mm	*in*		*mm*	*in*	
150	6		40	$1\frac{1}{2}$	
125	5		32	$1\frac{1}{4}$	
100	4		25	1	
80	3		20	$\frac{3}{4}$	
65	$2\frac{1}{2}$		15	$\frac{1}{2}$	
50	2		10	$\frac{3}{8}$	

Bends

Size		Price	Size		Price
mm	in		mm	in	
150	6		40	$1\frac{1}{2}$	
125	5		32	$1\frac{1}{4}$	
100	4		25	1	
80	3		20	$\frac{3}{4}$	
65	$2\frac{1}{2}$		15	$\frac{1}{2}$	
50	2		10	$\frac{3}{8}$	

Elbows

Size		Price	Size		Price
mm	in		mm	in	
150	6		40	$1\frac{1}{2}$	
125	5		32	$1\frac{1}{4}$	
100	4		25	1	
80	3		20	$\frac{3}{4}$	
65	$2\frac{1}{2}$		15	$\frac{1}{2}$	
50	2		10	$\frac{3}{8}$	

Elbows. (*a*) M.I.; (*b*) Weldable.

Size		Price		Size		Price	
mm	in	(*a*)	(*b*)	mm	in	(*a*)	(*b*)
150	6			40	$1\frac{1}{2}$		
125	5			32	$1\frac{1}{4}$		
100	4			25	1		
80	3			20	$\frac{3}{4}$		
65	$2\frac{1}{2}$			15	$\frac{1}{2}$		
50	2			10	$\frac{3}{8}$		

Flanges (*per pair*)

Pipe size		Price	Pipe size		Price
mm	in		mm	in	
150	6		80	3	
125	5		65	$2\frac{1}{2}$	
100	4		50	2	

Unions. (*a*) Malleable; (*b*) Capillary.

Size		Price		Size		Price	
mm	in	(*a*)	(*b*)	mm	in	(*a*)	(*b*)
50	2			20	$\frac{3}{4}$		
40	$1\frac{1}{2}$			15	$\frac{1}{2}$		
32	$1\frac{1}{4}$			10	$\frac{3}{8}$		
25	1						

Reducing sockets. M.I.

Larger size		Price	Larger size		Price
mm	in		mm	in	
65	$2\frac{1}{2}$		25	1	
50	2		20	$\frac{3}{4}$	
40	$1\frac{1}{2}$		15	$\frac{1}{2}$	
32	$1\frac{1}{4}$		10	$\frac{3}{8}$	

Water valves (fullway), weights as specified.

Size		Price	Size		Price
mm	in		mm	in	
150	6		40	$1\frac{1}{2}$	
125	5		32	$1\frac{1}{4}$	
100	4		25	1	
80	3		20	$\frac{3}{4}$	
65	$2\frac{1}{2}$		15	$\frac{1}{2}$	
50	2		10	$\frac{3}{8}$	

Radiator valves (as specified). Allow 0·7 m ($1\frac{1}{2}$ ft) of pipe for connection to main as basis of price. (*a*) Straight; (*b*) Angle

Size		Price		Size		Price	
mm	in	(*a*)	(*b*)	mm	in	(*a*)	(*b*)
32	$1\frac{1}{4}$			20	$\frac{3}{4}$		
25	1			15	$\frac{1}{2}$		
				10	$\frac{3}{8}$		

Empty gland cocks with hose unions.

Size		Price	Size		Price
mm	in		mm	in	
50	2		25	1	
40	$1\frac{1}{2}$		20	$\frac{3}{4}$	
32	$1\frac{1}{4}$		15	$\frac{1}{2}$	
			10	$\frac{3}{8}$	

Stopcocks (as specified).

Size		Price	Size		Price
32	$1\frac{1}{4}$		20	$\frac{3}{4}$	
25	1		15	$\frac{1}{2}$	

Air cocks.

Size		Price
mm	in	
6	$\frac{1}{4}$	
3	$\frac{1}{8}$	

Sleeves (*as specified*). To sleeve the following pipe sizes:

Size		Price	Size		Price
mm	in		mm	in	
150	6		40	$1\frac{1}{2}$	
125	5		32	$1\frac{1}{4}$	
100	4		25	1	
80	3		20	$\frac{3}{4}$	
65	$2\frac{1}{2}$		15	$\frac{1}{2}$	
50	2		10	$\frac{3}{8}$	

C. *Radiators.* Per square metre fixed complete with top stay, cantilever brackets and with valve unions and air cock fitted. Type of radiation according to specification.

Type	Price				
	330 *mm*	460 *mm*	610 *mm*	760 *mm*	910 *mm*
Steel or C.I. Hospital 180 mm					
,, ,, ,, ,, 127 mm					
,, ,, ,, ,, 76 mm					
,, ,, ,, 2 col.					
,, ,, ,, 3 col.					
,, ,, ,, 4 col.					
,, ,, ,, 6 col.					
,, ,, ,, Wall.					
Steel panel type.					

D. *Smoke pipe and fittings.*
Smoke pipe per mitre, supplied and fixed.

Size		Price	Size		Price
mm	in		mm	in	
350	14		150	6	
300	12		115	$4\frac{1}{2}$	
250	10		100	4	
200	8				

Smoke elbows supplied and fixed (with soot door)—90°

Size		Price		Size		Price
mm	*in*			*mm*	*in*	
350	14			150	6	
300	12			115	4½	
250	10			100	4	
200	8					

E. Thermal insulation (as specified), unit prices.

Note: Schedules similar to the above, embracing pipes and fittings in copper, stainless steel, and plastics, also convectors, heating panels, skirting heaters, etc. may be included as necessary.

THE FORM OF TENDER

Public Undertakings, Government Departments, Local Authorities, and large commercial organisations receive tenders on their own printed form. This is completed by the contractor who states the amount of his tender, the name and address of his surety, and the amount of the bond. The form, which must be signed, witnessed, and the Common Seal of the company affixed, also embodies the contractor's acknowledgement that his tender is based upon the drawings and specifications, and also that he has read, and accepts, the Terms and Conditions of the Contract.

The Form of Tender submitted by the successful contractor becomes a legal contract document. Such forms are not always used, and many tenders are submitted as normal business letters in which the contractor quotes a price based upon the owner's drawings and specifications. Such a document, properly signed and stamped, is just as legally binding as the pro forma type of tender document.

For contract work involving sums exceeding a few hundred pounds, most organisations prefer to have an agreement, or engrossment of the contract, legally drawn up and executed by both parties.

The documents to be dealt with by the heating contractor in preparing his tender have now been discussed, and these can be summarised as follows:

(1) Letter of 'invitation to tender'.
(2) Set of drawings.
(3) Standard Specification.
(4) Supplementary Specification.
(5) Variation Schedule, or a priced Bill of Quantities.
(6) Terms and Conditions of Tender, and Contract.
(7) The Form of Tender.

Heating Sub-contract

The Royal Institute of British Architects' 'Form of Main Contract' is now generally adopted for building works.

A standard form of Sub-contract (issued under the sanction of and approved by the National Federation of Building Trades Employers, and the Federation of Associations of Specialists and Sub-Contractors, and approved by the Committee of Associations of Specialist Engineering contractors) is available for use where the Sub-Contractor is nominated under the R.I.B.A. form of main contract.

Standard supplementary conditions for use with heating and ventilating tenders are available in booklet form, entitled *Standard Form of Conditions of Tender, for use in connection with Mechanical Engineering Systems in Buildings.* (See list of recommended publications in Appendix A.)

3

MATERIALS

Metrication of Sizes

Production of an entirely new range of metric heating and ventilating equipment, components, and materials will take some years to complete.

As an interim measure where the metric sized materials are not yet available, the manufacturers are adding the equivalent metric sizes to the Imperial dimensions shown in trade publications.

Light gauge copper tubes used for domestic water installations, formerly covered by British Standards 659, 1386, and 3931, are now contained in one metric standard, B.S. 2871: Part 1, in which the tube sizes refer to outside diameter measurements in millimetres. Soft temper copper tube for use with microbore heating is manufactured to B.S. 2871: Part 2. By multiplying the Imperial nominal bore by 25·4, and contracting the results to the nearest whole number we get the translation shown in line 2, with the relative B.S. 2871 outside diameters in line 3:

Copper tubes B.S.2871

Imperial nominal bore	$\frac{1}{4}$ in	$\frac{3}{8}$ in	$\frac{1}{2}$ in	$\frac{3}{4}$ in	1 in	$1\frac{1}{4}$ in	$1\frac{1}{2}$ in	2 in
Translated metric nominal bore	6 mm	10 mm	13 mm	19 mm	25 mm	32 mm	38 mm	51 mm
Metric o.d. size B.S.2871	8 mm	12 mm	15 mm	22 mm	28 mm	35 mm	42 mm	54 mm

Standard metric sizes for steel tubes have been agreed, and are covered by I.S.O. Recommendation R65. The recommended metric sizes are the exact equivalents of the Imperial sizes in B.S. 1387, and are completely interchangeable.

Metric and Imperial sizes up to 2 in nominal bore are as follows:

Mild steel tubes B.S.1387

Metric nominal bore	10 mm	15 mm	20 mm	25 mm	32 mm	40 mm	50 mm
Imperial nominal bore	$\frac{3}{8}$ in	$\frac{1}{2}$ in	$\frac{3}{4}$ in	1 in	$1\frac{1}{4}$ in	$1\frac{1}{2}$ in	2 in

On receipt of the invitation to tender, the contractor will post an acknowledgment, stating his willingness to tender, or otherwise. If the decision is to submit a tender, the drawings and specifications and information as to the latest date for submission of the tender, are passed to the estimator for pricing.

The estimator will make a careful study of the drawings and specifications to familiarise himself with the scheme, layout of plant, sizes and quality of equipment and materials. With the drawings and specifications before him he is now in a position to take off the quantities and prepare a list for pricing. Although most of the materials can be priced from the manufacturers' published price sheets, items will occur which are not price listed and for which special quotations will have to be obtained.

Those materials in constant demand which can be priced from the catalogues and Trade Discount sheets, are listed below.

Boilers and Boiler Gear (Heating and H.W.S.)

Boilers (complete).
Safety valves.
Empty cocks.
Smoke pipe.
Smoke bends.
Smoke hood.
Soot doors and frames.
Draught stabilisers.

Altitude gauges.
Thermometers.
Boiler thermostats.
Damper regulators.
Boiler flow and return headers.
Boiler pipe fittings and flanges.
Stoking tools and racks.
Boiler insulating jackets.

Pipes and Fittings

Pipes of cast iron, steel, aluminium, and copper.

Pipe fittings in cast iron, malleable iron, wrought iron, in all sizes and of the following types, which may be screwed, flanged, or socketed:

Elbows, equal and reducing, 90° and 45°.
Bends and springs, male and female, 90° and 45°.
Tees, equal or reducing. Tongue or plain saddle tees.
Cross tees, equal or reducing. Tongue or plain. Y-branches.
Sockets, parallel, taper, or right- and left-handed thread.

Sockets, reducing.   Twin elbows.
Blanking sockets.   Expansion joints.
All standard pipe fittings, welding fittings, elbows, bends, and
    branch ends.
Jointing materials.
Joint rings and packing.
Copper and aluminium fittings of all types.
Polythene tubes and all fittings for cold water.

Pipe supports, ring hangers and plates, hooks and straps, roller
chairs and brackets, cradles and clamps, wall brackets, schoolboard brackets, and girder lugs.

Valves in Bronze or Cast Iron

Globe wheel valves.
Gate wheel valves.
Quick opening valves, lever handle or pedal operated.
Angle valves.
Needle point valves.
Check valves.
Featherweight back pressure valves.
Radiator valves, wheel, lockshield, or closed shield pattern.
Throttle valves.
Bronze cocks.
Air valves.

Radiators and Other Heaters

Cast iron and steel radiators of heights up to 910 mm (36 in) and
    widths up to 220 mm ($8\frac{5}{8}$ in) and 340 mm ($13\frac{1}{2}$ in) for window
    pattern radiators, wall type.
Radiators in brass, copper, and aluminium.
Radiator baffle plates.
Radiator supports, legs, top stays; skirting heating panels.
Radiator valves.
Radiator unions.
Radiator air cocks and keys, automatic air valves.
Radiator wall shields.
Radiator saddles.
Radiator fresh air inlets, baffle plates and operating gear.
Convectors, radiant panels.
Unit heaters, pipe panels for concealed radiant heating.
Skirting and heating panels.

Tanks and Cisterns for Water or Oil

Open steel tanks and cisterns of all sizes, galvanised or ungalvanised.
Ball valves of all sizes.
Tank brackets and supports.
Liquid level gauges.
Gauge glasses and cocks.
Metal tank lids or covers.

Cylinders and Calorifiers for Hot Water Supply

Direct and indirect cylinders from 91 to 910 litres (20 to 200 gallons) capacity, in galvanised mild steel or copper for use with low pressure hot water.
Small steam calorifiers for high and medium pressures.
Water gauges, pressure gauges, thermometers, safety valves.
Steam and water thermostatic control valves.
Back pressure valves.
Steam traps.
Strainers.
Automatic air release valves.
Towel rails, and supports.

Sump Pumps

Hand-operated semi-rotary pumps, in iron, brass, or bronze.
Electrically-operated centrifugal drainage and sump pumps.
Foot valves.
Strainers.
Electrical starter and switchgear.
Float switches.

Pumps and Accelerators (Belt and Direct Driven)

Pumps and accelerators of listed sizes and capacities.
Starters and electrical control gear.
Accelerators and control equipment for small-bore heating systems.
Pump counter flanges.
Lubricators.
Bronze body pumps for hot water supply.
Pump air cocks.
Pump drain cocks.

Ventilating Equipment

Window fans.
Propeller fans.
Outlet bends, automatic shutters, outlet cowls.
Axial flow fans.
Air filters.
Air heaters.
Air diffusers.
Air grilles and other air distributing equipment.
Roof extractor ventilators (natural).
Variable opening ventilators (natural).
Cowl type ventilators.
Electrical switchgear and controls.
Canvas sleeve connectors.
Dampers.
Anti-vibration mountings.
Air duct silencers.
Fume cupboard fans.
Plastic ducts, tubes, and fittings up to 450 mm (18 in) diameter.
Lightweight fibre ducts and tubes up to 910 mm (36 in) diameter.
    Standard length 4·57 m (15 ft).
Black and galvanised sheet steel ducts and fittings.
Sheet copper and aluminium ducts and fittings.
Prefabricated plaster ducts and fittings.
Asbestos cement ducts and fittings.

Special Quotations

Special quotations from the manufacturers are often necessary for equipment to fulfil duties laid down in the Supplementary Specification, and for prefabricated pipework, and special pieces such as large welded manifolds and flue headers as indicated on the drawings. Equipment and materials under this heading include:

Automatic coal stokers.
Oil burners.
Large radiant heating panels.
Pumps.
Steam boilers.
Boiler room instruments and control panels.
Steam feed pumps.
Pressurising equipment.
Special flue connections and headers.

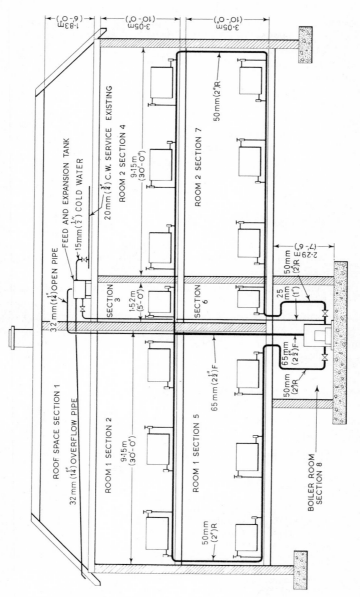

Figure 5. Low pressure hot water system in a two-storey building

Special pipe and flue supports, hangers and brackets.
Special pipe manifolds, assemblies and large expansion loops.
Special pipe sections and fabrications.
Ventilating fans.
Air washers and humidifiers.
Refrigerating plant.
Large air heaters.
Special controls as specified.
Induced draught fans.
All plastic ventilating fans.
Plastic tapers, branches, bends, and dampers.

TAKING OFF THE QUANTITIES

When taking off the quantities the job should be divided into convenient sections, each of which is dealt with separately. The method is made clear by reference to Figure 5, which represents a low pressure gravity hot water heating system in a two-storey building. The procedure is to start at the top of the building and work downwards.

Every care is taken to eliminate error when taking off, and sectionalising the work makes it much easier to do this. The cost of each section of the work can be seen at a glance, and the Bill of Quantities becomes doubly useful when costing work during erection, and also when costing applications for interim payments as the job proceeds.

LIST OF MATERIALS AS MEASURED FROM FIGURE 5

SECTION 1. ROOF SPACE

No.	Description
1	45 litre expansion tank, galvanised.
2	tank supports.
1	15 mm ball valve.
10·36	metres of 32 mm galvanised pipe for overflow.
1	32 mm–45° galvanised bend for overflow outlet.
3	metres of 15 mm galvanised pipe (CW to tank).
1	15 mm stop cock (CW to tank).
1	15 mm galvanised bend.
1	20 × 20 × 15 mm galvanised tee.
1	15 mm galvanised nipple.
1	15 mm union.
1·83	metres of 25 mm galvanised pipe, for feed and expansion pipe.
1	bronze gate valve.
1	25 mm nipple.

SECTION 1. ROOF SPACE (*Contd.*)

No.	Description
1	25 mm bend.
1	25 mm pipe sleeve.
3	metres of 32 mm galvanised pipe for open vent.
2	32 mm galvanised bends.
1	32 mm pipe sleeve.
5	32 mm pipe brackets. Screw-on type.
1	25 mm pipe bracket. Screw-on type.
2	15 mm pipe brackets. Screw-on type.
1·83	metres of insulation for feed and expansion pipes. Standard Specification, Section (e2).
3	metres of insulation for cold water pipe to feed and expansion tank, as Standard Specification, Section (e2).

SECTION 2. ROOM 1, FIRST FLOOR

No.	Description
9·14	metres of 50 mm M.S. pipe.
1	50 mm bend.
1	50 mm union.
1	$50 \times 50 \times 65 \times 32$ mm pitcher cross.
1	65 mm pipe sleeve.
1	65 mm floor plate.
1	50 mm pipe sleeve.
1	50 mm floor plate.
1	50 mm wall plate.
6	$50 \times 50 \times 25$ mm tongue tees.
3	50 mm build-in schoolboard pipe brackets.
3·66	metres of 25 mm M.S. pipe for radiator connections.
3	25 mm easyclean bronze radiator angle wheel valves.
3	25 mm easyclean bronze radiator lockshield valves.
3	radiator top stays.
6	radiator build-in wall brackets (2 per radiator).
3	915 mm high, 146 mm wide, 20 section hospital radiators, each 5·57 m^2.
3	radiator air cocks.
3	metres of 32 mm galvanised pipe (open vent).
1	32 mm build-in schoolboard pipe bracket.
3	metres of 25 mm galvanised pipe for feed and expansion pipe.
1	25 mm build-in schoolboard pipe bracket.
1	25 mm ceiling plate.
1	32 mm ceiling plate.

SECTION 3. CORRIDOR, FIRST FLOOR

No.	Description
1	50 mm union.
2·13	metres of 50 mm M.S. pipe.
2	50 × 50 × 25 mm tongue tees.
1	50 mm pipe sleeve.
2	50 mm wall plates.
1·22	metres of 25 mm M.S. pipe for radiator connections.
1	25 mm easyclean bronze angle radiator wheel valve.
1	25 mm easyclean bronze angle radiator lockshield valve.
2	radiator build-in wall brackets.
1	radiator top stay.
1	915 mm high, 146 mm wide, 13 sections, 3·62 m² hospital radiator.
1	radiator air cock.

SECTION 4. ROOM 2, FIRST FLOOR

No.	Description
1	50 mm union.
9·14	metres of 50 mm M.S. pipe.
1	50 mm bend.
3	50 mm build-in schoolboard pipe brackets.
2	50 mm pipe sleeves.
1	50 mm wall plate.
1	50 mm floor plate.
6	50 × 50 × 25 mm tongue tees.
3·66	metres of 25 mm M.S. pipe for radiator connections.
3	25 mm easyclean bronze radiator angle wheel valves.
3	25 mm easyclean bronze radiator angle lockshield valves.
3	radiator top stays.
6	radiator build-in wall brackets.
3	915 mm high, 146 mm wide, 20 section, hospital radiators, each 5·57 m².
3	radiator air cocks.

Pipework

Pipework is measured from the plans and elevations and listed according to size in descending order of magnitude.

When measuring pipework add 5 per cent to allow for waste in cutting and screwing.

Pipe brackets and supports can be measured with the pipes. This can only be done accurately by the estimator marking the position of each pipe bracket on the drawings. An approximate method, not recommended for accurate estimating, is to divide the length of pipe as measured by the spacing required in Clause 23, Section A, of the Standard Specification.

The qualities of tube to be taken for are generally as stated in Clause 20, Section A, of the Standard Specification.

An important point to watch when taking off pipework is the scale of the drawing used for measurement. Costly mistakes have been made due to mis-reading of the scale by estimators when working under pressure.

SECTION 5, ROOM 1, GROUND FLOOR

No.	Description
11·58	metres of 50 mm M.S. pipe.
2	50 mm bends.
4	50 mm build-in pipe brackets.
1	50 mm floor sleeve.
1	50 mm ceiling plate.
1	50 mm floor plate.
1	50 mm union.
6	50 × 50 × 25 mm tongue tees.
3·66	metres of 25 mm M.S. pipe.
3	25 mm easyclean bronze radiator angle wheel valves.
3	25 mm easyclean bronze radiator angle lockshield valves.
3	radiator top stays.
6	radiator build-in brackets.
3	915 mm high, 146 mm wide, 16 section, hospital radiators, each 4·46 m².
3	radiator air cocks.
3·35	metres of 65 mm M.S. pipe.
1	65 mm build-in schoolboard pipe bracket.
1	65 mm pipe sleeve.
1	65 mm floor plate.
1	65 mm ceiling plate.
3·35	metres of 25 mm galvanised pipe (feed and expansion pipe).
1	25 mm build-in schoolboard pipe bracket.
1	25 mm pipe sleeve.
1	25 mm floor plate.
1	25 mm ceiling plate.

SECTION 6. CORRIDOR, GROUND FLOOR

No.	Description
1	50 mm union.
1·83	metres of 50 mm M.S. pipe.
1	50 mm build-in schoolboard pipe bracket.
1·83	metres of 50 mm M.S. pipe.
1	50 mm pipe sleeve.
1	50 mm floor plate.
1	50 mm wall plate.
1	50 mm bend.
2	50 × 50 × 25 mm tongue tees.
1·22	metres of 25 mm M.S. pipe (radiator connections).
1	25 mm easyclean bronze angle radiator wheel valve.
1	25 mm easyclean bronze angle radiator lockshield valve.
2	radiator build-in wall brackets.
1	radiator top stay.
1	915 mm high, 146 mm wide, 10 section hsopital radiator, 2·79 m².
1	radiator air cock.

Pipe Fittings

Pipe fittings can now be 'taken off' and again the estimator will refer to the Specification, which defines the types of joints and connections allowed for various sizes and types of tube, and other requirements which may affect the cost of the fittings and the labour required to install.

Sleeves occur where pipes pass through floors and walls and can easily be counted from the plans. The Supplementary Specification should be consulted for the type of sleeve called for, as this will of course affect the price per sleeve.

SECTION 7. ROOM 2. GROUND FLOOR

No.	Description
11·58	metres of 50 mm M.S. pipe.
4	50 mm build-in schoolboard pipe brackets.
1	50 mm wall sleeve.
1	50 mm wall plate.
1	50 mm ceiling plate.
6	50 × 50 × 25 mm tongue tees.
3·66	metres of 25 mm M.S. pipe (radiator connections).
3	25 mm easyclean bronze radiator angle wheel valves.
3	25 mm easyclean bronze radiator angle lockshield valves.
3	radiator top stays.
6	radiator build-in wall brackets.
3	915 mm high, 146 mm wide, 15 section hospital radiators, each 4·18 m².
3	radiator air cocks.

Radiators

Radiators, radiator valves, pipe connections, top stays, wall brackets, air valves, and shelves, should be taken off together, and are easily measured from the radiator schedule. This applies to other forms of heater, such as unit heaters, radiant panels, skirting heaters, convectors, and concealed panels.

Where outside air grilles, inside adjustable registers and wall shields are included for radiator installations, they should be grouped with the radiator when taking off, and when pricing the Bill.

Control Valves and Empty Cocks

Stopvalves, wheel or lockshield, and also empty cocks, controlling main and branch circuits, should be taken off with the pipework in each section.

In Figure 5 all pipes, with the exception of the boiler room, are above ground. In many schemes, however, much of the distribution pipework from the boiler room to the various sections of the building is run in subways and trenches below the ground floor. When taking off in such cases the whole of the pipework in main subway or trench is grouped into one section. Pipes in small trenches, chases, and ducts, in individual rooms, are grouped with those rooms.

Boiler Plant

Particulars of the boiler or boilers are taken from the Specification, and the boiler output, heating surface, boiler mountings, tappings, controls, smoke connections, and insulation should be very carefully listed.

The boiler room is regarded as a separate section for estimating purposes, and all plant, valves, pipework, pumps, cylinders, and all equipment to be fixed in the boiler room should be grouped together.

Cylinders and Calorifiers

Manufacturers require the following information when considering an enquiry for hot water cylinders and steam calorifiers:

(1) Address of site or building.
(2) Static head of system.
(3) Hot water storage capacity of the vessel.
(4) Storage temperature.

SECTION 8. BOILER ROOM

No.	Description
1·83	metres of 65 mm M.S. pipe.
2	65 mm bends.
2	65 mm unions.
8·53	metres of 50 mm M.S. pipe.
1	50 × 50 × 25 mm tee.
2	50 mm nipples.
4	50 mm build-in schoolboard galvanised pipe brackets.
8	50 mm bends.
2	50 mm bronze lockshield gate valves.
3·66	metres of 25 mm galvanised pipe.
3	25 mm bends (cold feed and expansion pipe).
1	25 mm union.
2	25 mm build-in schoolboard pipe brackets.
1	Cast iron sectional hot water heating boiler, with water cooled grate, for hand firing, having 6 sections, total heating surface 3·30 m². Output 45·72 kW, complete with smoke hood, doors, and dampers. Screwed tappings, 1–80 mm flow, and 2–80 mm returns.
1	80 mm to 65 mm bush for flow connections.
2	80 mm to 50 mm bushes for return connections.
1	galvanised steel insulating jacket.
1·83	metres of 200 mm cast iron smoke pipe, self colour.
1	smoke pipe sleeve.
1	200 mm smoke bend, 112½°, self colour.
1	set of stoking tools, and rack.
1	20 mm totally enclosed deadweight safety valve.
1·83	metres of M.S. pipe for safety valve discharge.
1	10 mm bend.
1	120 mm dial Bourdon pattern altitude gauge.
1	203 mm long brass cased thermometer with revolving shield, screwed 15 mm, temperature range 4°C to 120°C (40°F to 240°F).
1	thermometer pocket, screwed 25 mm male thread, and 15 mm female thread. Length of tail, 63 mm.
1	25 mm gland pattern bronze empty cock, screwed female, with hose union and key.
1	25 mm nipple for fitting empty cock to boiler tapping.
1	damper regulator, screwed 50 mm, with malleable iron well.
1	225 × 150 mm soot door and frame for building into brick flue by builder.
1	blue brick boiler base by builder, formed with 112½ mm bull-nosed brick on edge with 225 × 225 mm corner blocks.
	Insulate to Standard Specification (e1)
1·52	metres of 65 mm pipe.
2	65 mm bends.
6·10	metres of 50 mm pipe.
8	50 mm bends.

(5) Primary hot water temperature (if indirect cylinder).
(6) Whether primary circulation is to be gravity or pump operated.
(7) Sketch of tappings. Position of manholes, surface of heater battery.
(8) Metal required, galvanised steel, or copper.
(9) Steam pressure (for calorifiers).
(10) Steel or copper heater.
(11) Chemical analysis of water.
(12) Date when delivery will be expected.

With regard to item (11), the chemical composition of the water may affect the output of a steam calorifier, and for this reason it is advisable that the manufacturer should see a chemical analysis when quoting for the equipment.

Tanks and Connections

As with boilers, particular note is made of the number, size, position, and type of tappings. Usually there is one inlet controlled by a ball valve. There may, however, be more than one outlet to cater for, together with the control valves so necessary for maintenance purposes. Where multiple tank arrangements are shown, any interconnections between the tanks, and also the proposed method of dealing with the overflow from each tank, will be carefully noted.

Special quotations for tanks are often needed from the manufacturer, and the particulars referred to above must be clearly stated in the contractor's enquiry, together with the metal (cast iron or steel), gauge of metal for tanks and tank covers, and whether to be galvanised or not.

Tank supports are usually provided and fixed by the builder under a separate contract. The estimator will check this point with the Specification, and if supports are the responsibility of the heating contractor, he will allow for these in the Bill.

Pipework in the tank room, or in the roof space where the tanks are to be fitted, including all valves, fittings, and supports for cold water supply to tanks, overflow pipes, and balancing pipes is included in this section of the work.

Outlet pipes, open vents, and the cold water supply pipe are measured from the tanks to the position where they leave the tank room or roof space. Where these pipes pass through other rooms en route to the boiler room or other parts of the building, they are included in the quantities for these rooms.

Thermal Insulation

The boilers, cylinders, pipes, and other surfaces to be insulated, are carefully measured from the drawings, and the Specification consulted as to type and thickness of covering required.

Unless the heating contractor employs skilled coverers, as some firms do, the estimator will obtain a quotation from a specialist firm.

Before submitting his price the insulating contractor should if possible examine the drawings and specification in order that he may know exactly the conditions under which his men will work. This applies especially to covering in ducts and trenches. Although many such quotations are based on a schedule of equipment, pipe lengths and sizes, a much closer estimate will result from the covering contractor's examination of the drawings.

If the quotation is invited on a schedule basis, a detailed description of the boilers, cylinders, pipes, and fittings to be covered, and also a full specification of the insulation, must accompany the enquiry.

Most heating contractors will have sufficiently up-to-date price knowledge of insulation to enable them to prepare an estimate for normal everyday work, without sending an enquiry to a covering firm every time they need a price. When estimates are prepared in this way the estimator must satisfy himself that the work he is estimating for is really comparable with the past work or any current work upon which his prices are based. Factors like height of pipes, accessibility, and any special requirement of the Specification concerning type, quality, thickness, or finish which may require pricing by a specialist, should be noted.

Pumps

The price to include for the circulating pumps will invariably be based on a maker's quotation. When sending the enquiry full information on the following points will be needed by the maker:

(1) Number of pumps required, and address of site to which they are to be delivered.
(2) Operating conditions:

> Litres per second to be circulated.
> Total pressure in bars against which the pump operates.

(3) Diameter of connections, and whether screwed or flanged.
(4) Details of drive:

Type of drive, direct-coupled, belt-driven. Prime mover.
Electric motor, petrol engine, oil engine, or steam turbine.
If electric motor state whether a.c. or d.c. and voltage,
   phases, frequency.
Type of motor enclosure.

(5) Any special requirements as to connections, mountings,
   silencer, etc.

Fittings such as drain cocks and drip pipes, isolating valves, special
evase connections between pumps and piping systems and head
gauges as specified will be included.

The electrical starters will be included in the maker's quotation.
Wiring to the pump starters and motors must be allowed for, and
a price obtained from an electrical contractor, which the estimator
will include in his Bill.

Automatic Coal Stokers and Oil Burners

These items are often covered in the Specification by Provisional
Sums to be included in the tender. The machines may, however, be
specified to be provided and fixed by the heating contractor, in
which case the estimator obtains prices based on the supply and
installation of machines and control equipment as laid down in the
Standard and Supplementary Specifications.

Heating firms do not as a rule employ automatic stoker mechanics
and an enquiry must therefore go to a stoker manufacturer, who
will require the following information:

(1) Address of site.
(2) Supply and fixing of stoker and controls.
(3) Rating of stoker (as Supplementary Specification).
(4) Wiring work (as Standard Specification).
(5) Characteristics of available electricity supply.

All materials for the automatic stoker installation will be supplied
by the maker of the machine, including electrical control equipment
and thermostats, which he knows from experience to be most suitable for the efficient operation of his machines.

The isolating switch fuse and stoker wiring materials to be fixed
in the boiler room are normally provided by the electrical contractor,
who quotes the heating contractor for this work. Wiring diagrams
provided by the stoker maker must accompany the heating contractor's enquiry to the electrical firm. If extra costs are to be avoided
during the erection stage it is essential for the electrical contractor

to be provided with complete and detailed information upon which to base his price.

The above remarks, as applied to the machine, its electrical and thermostatic controls and ancillary wiring, also hold good for oil-burning installations.

The installation of oil storage tanks, filler pipe, open vent pipe, sludging pipe, outlet pipe from tank to burner, tank oil gauge, and all valves and fittings necessary for storing and conveying the oil to the burner, will be carried out by the heating contractor, and therefore all materials for this section of the work are taken off in the normal way.

Provisional Sums may be included in the engineering specification by the architect or the consulting engineer, for wiring and other work associated with the heating or ventilating installation. These amounts are included in the engineering tender, and no action by the heating contractor is necessary until the contract is let, when estimates for the works covered by these items will be invited, and sub-contractors appointed.

VENTILATION PLANT AND EQUIPMENT

As with heating and hot water work, careful perusal of both specification and drawings is necessary when taking off ventilating quantities.

Large pieces of special equipment like fans, filters, air washers and heaters are usually subject to special quotations and are taken off as single items, to be inserted in the Bill when the manufacturer's estimate is received. When inviting quotations for this type of plant the manufacturer should have a copy of the specifications and scheme drawings, together with the technical design data such as resistance and air volumes upon which the consultant's or the contractor's plant sizing is based. Special requirements regarding multi-speed fans, noise limits, automatic controls, air temperatures and pressure must be freely available to enable the manufacturers to put forward the correct plant for the job.

Ventilating equipment is extremely costly and the estimator who may not be dealing with tenders for such plant every day is well advised to seek specialist guidance in the preparation of estimates of this nature.

These remarks apply equally to mechanical and natural ventilation work.

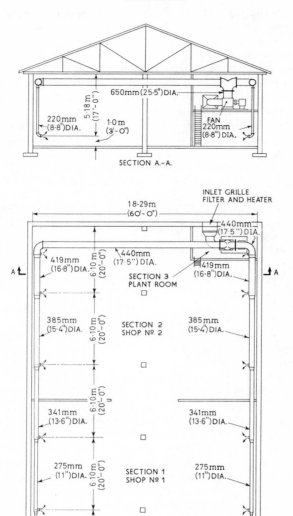

Figure 6. Plenum ventilating system for two factory workshops

When pricing for natural ventilation work, the manufacturer's advice should always be sought regarding the efficiencies of roof ventilators and other equipment which rely mainly upon wind effect for their proper functioning.

The process of taking off for ventilating work can be sectionalised in the manner described for heating. As already stated, centrifugal fans and other large items are included as separate items in the Bill, and priced on the maker's quotation.

Ductwork, including brackets, joints, and dampers, is measured from the drawings listed in the Bill according to size, and priced by weight as delivered to the site ready for erection, the unit employed being *price per tonne*. This will of course vary, like most commodities, according to the price of raw materials and labour. In the case of ductwork a further variable operates, namely the simplicity or otherwise of the design and shape of the ducting. At the present time the price of galvanised sheet metal ducting for the normal type of ventilating varies, according to size and location of job, between £450 and £500 per tonne, plus the cost of erection at £125 to £150 per tonne.

The measured ductwork is listed and priced according to weight. The weight is calculated by reference to the Standard Specification, Section B, Clause 21, and a sheet metal weight table, as described later.

Other items include brackets and supports, flexible connections, grilles, and registers, machine mountings, electrical wiring, and equipment.

Figure 6 shows a plenum ventilating system installed to heat and ventilate two factory workshops. The plant consists of a centrifugal fan, dry filter, and an air heater, mounted on a high level platform at one corner of shop No. 2. The heater receives a steam supply from the factory system, and condensate is piped from the heater to the nearest condensate main and returned to the factory boiler house. Warm air is delivered to the workshop from the fan outlet by two lines of galvanised steel ducts, run overhead at a height of approximately 5·2 m (17 ft). Twelve downpipes convey the warm air from the main overhead duct, to discharge at a level of approximately 1·0 m (3 ft) from the shop floor. Each downpipe is 220 mm (9 in) diameter, fitted with a twin 220 mm (9 in) bend at the outlet point. Dampers or throttle pieces are fitted in the downpipes to balance the air delivery to all parts of the workshops. Fresh air is drawn from outside through a louvred inlet grille, to the filter, heater, and fan inlet.

The quantities are taken off, and listed in the following order:

LIST OF QUANTITIES

SECTION 1. WORKSHOP NO. 1

No.	Description
6	downpipes, each 4·6 m long × 220 mm diameter × 24 B.G.
6	bends at branches with duct.
6	twin bends at discharge points.
6	dampers.
12	fixing brackets for downpipes. 25 × 3 mm flat iron.
12·2	metres of 220 mm diameter × 24 B.G. duct at high level.
6	flat iron brackets for above duct. 30 × 4 mm.
75	10 mm diameter × 50 mm Whitworth bolts and nuts.
150	10 mm washers.
12·2	metres 275 mm diameter × 24 B.G. duct with stiffeners at joints (at high level).
6	flat iron brackets. 25 × 3 mm.
25	10 mm diameter × 50 mm Whitworth bolts and nuts.
50	10 mm washers.
6·1	metres of 341 mm diameter × 24 B.G. with stiffeners at joints (at high level).
4	flat iron brackets. 25 × 3 mm.
16	10 mm diameter × 50 mm Whitworth bolts and nuts.
32	10 mm washers.

SECTION 2. WORKSHOP NO. 2

No.	Description
6	downpipes, each 4·6 m long × 220 mm diameter × 24 B.G. (0·6 mm).
6	bends at branches of main duct.
6	twin bends at discharge points.
6	dampers.
12	downpipe fixing brackets, 30 × 3 mm flat iron.
6·1	metres of 341 mm diameter × 24 B.G. duct with stiffeners (at high level).
4	flat iron brackets, 30 × 3 mm.
12·2	metres of 385 mm diameter × 24 B.G. duct with stiffeners (at high level).
6	flat iron brackets, 30 × 3 mm.
5·2	metres of 419 mm diameter × 24 B.G. duct with stiffeners (at high level).
7	flat iron brackets, 30 × 3 mm.
14·94	metres of 440 mm diameter × 24 B.G. duct with stiffeners (at high level).
6	flat iron brackets, 40 × 3 mm.
170	10 mm diameter × 50 mm Whitworth bolts and nuts.
340	10 mm washers.

Calculating the Weight of Ductwork

Careful measurement of ductwork from the drawings is essential when taking off ventilation quantities. Plans, elevations, and sectional elevations of the system are necessary.

Bends and branch connections need to be measured accurately, and those parts of the ducting which actually pass through walls, or which may be hidden by masonry or other parts of the building structure and are not clearly shown on 1:100 scale drawings, must not be missed. Where doubt exists the estimator may require to see a 1:20 detail of any part of the proposed installation not properly measurable on the 1:100 drawings.

SECTION 3. PLANT ROOM

No.	Description
1	440 × 440 × 650 mm × 22 B.G. (0·8 mm) twin bend at fan outlet.
1	1·22 m making-up piece between fan outlet and 650 mm diameter branch of twin bend.
2	canvas connections to fan.
2	flat iron straps in halves for fixing canvas connections to inlet and outlet ducts.
2	flange pieces for fixing canvas connections to fan inlet and outlet.
25	12·5 × 50 mm Whitworth bolts and nuts.
50	12·5 mm washers.
1	slow speed multivane centirfugal fan to deliver 3·54 m³/s against a total pressure of 3·11 mbar.
1	complete set of anti-vibration mountings and foundation bolts.
1	1050 mm wide × 1200 mm high, single row, copper gilled tube heater, arranged with 32 mm bore steam, and 25 mm bore condensate connections.
1	throw-away type wool air filter.
50	12·5 × 50 mm Whitworth bolts and nuts.
100	12·5 mm washers.
1	1050 × 1200 mm × 20 B.G. (1·0 mm) duct between fresh air inlet and heater.
1	1050 × 1200 mm inlet grill to be built into outside wall by builder.
1	duct thermostat in fan discharge duct.
1	32 mm thermostatic steam control valve for heater and 2 m of capillary tube to duct thermostat.
3	metres of 32 mm steam pipe.
2	32 mm pipe brackets.
1	32 mm steam stopvalve.
1	10 mm automatic air vent.
1	80 × 80 × 32 mm tee in steam main.
2	32 mm steam bends.
1	32 mm union.
6·1	metres of 25 mm condensate pipe.
4	25 mm pipe hangers.
1	25 mm steam trap.
1	25 mm pressure valve.
1	50 × 50 × 25 mm tee in condensate main.
6	25 mm steam bends.

SECTION 3. PLANT ROOM (*Contd.*)

No.	Description
2	25 mm equal tees for steam trap by-pass.
3	25 mm steam valves for steam trap isolation and by-pass.
2	25 mm elbows.
4	25 mm nipples.
1	25 mm union.
	a supply of graphite joint paste.
6·1	metres of glass silk insulation for 25 mm pipe as Standard Specification, Clause 38, Section (d2).
3	metres of glass silk insulation for 32 mm pipe as Standard Specification, Clause 38, Section (d2).
1	room thermostat.
1	triple pole and neutral, 15 A switch fuse to control fan motor circuit.
1	auto-contractor starter for fan motor. Wiring to fan isolator from nearest fuse board. Wiring to room thermostat, auto-contractor, and motor, from isolator.

The gauge or thickness of metal from which the ducts, fan, filter, and heater connections are made, is usually indicated on the drawings, in addition to the schedule in the Standard Specification. (See Section B, Clause 21.)

Having measured and listed the various sizes, the weight of the ductwork can be calculated by reference to Tables 2 and 3.

Table 3 gives the circumference of circles in metres, and the area in m² of sheet metal for circular ducts is found by multiplying column 2, 3, 5, or 6 by the measured length of duct.

Table 2. WEIGHT OF SHEET STEEL

(1) B.G.	(2) Thickness	(3) Black mild steel sheet		(4) Galvanised mild steel sheet		(5) Mild steel duct galvanised after manufacture	
	mm	kg/m²	lb/ft²	kg/m²	lb/ft²	kg/m²	lb/ft²
16	1·6	12·70	2·60	12·75	2·61	13·96	2·86
18	1·2	10·06	2·06	10·10	2·07	11·32	2·32
20	1·0	7·96	1·63	8·01	1·64	9·23	1·89
22	0·8	6·35	1·30	6·40	6·40	7·62	1·56
24	0·6	5·03	1·03	5·08	1·04	6·30	1·29

Table 3. CIRCUMFERENCE OF CIRCULAR DUCTS, AND PERIMETER OF
RECTANGULAR DUCTS
(Standard sizes to H.V.C.A.: DW/112)

(1) Duct size		(2) Rectangular duct perimeter m	(3) Circular duct circumference m	(4) Duct size		(5) Rectangular duct perimeter m	(6) Circular duct circumference m
Rectangular duct mm	Circular equivalent mm			Rectangular duct mm	Circular equivalent mm		
150 × 100	134	0·5	0·421				
250 × 100	171	0·7	0·537	500 × 300	424	1·6	1·332
200 × 150	190	0·7	0·597	400 × 400	440	1·6	1·382
250 × 150	212	0·8	0·666	700 × 300	496	2·0	1·559
200 × 200	220	0·8	0·691	600 × 400	537	2·0	1·688
400 × 150	264	1·1	0·829	700 × 400	578	2·2	1·816
300 × 200	269	1·0	0·845	600 × 500	603	2·2	1·894
250 × 250	275	1·0	0·864	700 × 500	650	2·4	2·042
300 × 250	301	1·1	0·946	700 × 600	713	2·6	2·241
500 × 200	341	1·4	1·072	700 × 700	761	2·8	2·391
600 × 200	371	1·6	1·166	800 × 600	770	2·3	2·419
500 × 250	385	1·5	1·210	800 × 700	824	3·0	2·589
600 × 250	419	1·7	1·316	800 × 800	881	3·2	2·832

Example. Find the weight of 45·72 m of 440 mm diameter duct made from 0·6 mm thick (24 B.G.) galvanised mild steel.

Circumference of duct, from Table 3 = 1·60 m
Area of duct = 1·60 m × 45·72 m
$\qquad$ = 73·15 m²
From Table 2, col. 4, weight of 1 m² of duct = 5·08 kg
then, weight of duct = 73·15 m² × 5·08 kg
$\qquad$ = 372 kg.
For estimating purposes add 10 per cent
$\qquad$ = 372 kg + 37 kg
$\qquad$ = 409 kg.

To obtain the true weight upon which to base an estimate, the weight of flanges, bolts, nuts, washers, bracings, stiffeners, brackets, supports, baffles, dampers, and throttle pieces must be added to the net weight of the duct. Special dampers and duct fittings, such as silencers, will be separate items in the Bill, and not included in the weight of the general ductwork for pricing purposes.

Flanges, stiffeners, brackets, and duct supports are generally made from flat iron or angle iron. Flanged joints may be of angle iron, or the duct ends may be turned up to form the flanges, between which a steel stiffening ring is fitted to give the joint strength. Stiffening rings are also fitted in the ducts to strengthen slip-on joints.

For estimating purposes add 10 per cent to the net weight to allow for overlapping at joints, for rivets, and for the scrap produced in fixing. This addition is also added to the materials used for supports and brackets made on site.

Table 4 gives the weight per foot for angle, channel, and flat bars, of the sizes commonly used for the construction and erection of ductwork.

Table 4. WEIGHTS OF STEEL ANGLES, FLAT BARS AND CHANNELS

Steel angles size		Unit length weight		Flat bars size		Unit length weight	
mm	in	kg/m	lb/ft	mm	in	kg/m	lb/ft
25 × 25 × 3	1 × 1 × $\frac{1}{8}$	1·19	0·80	25 × 3	1 × $\frac{1}{8}$	0·625	0·42
30 × 30 × 3	1$\frac{1}{4}$ × 1$\frac{1}{4}$ × $\frac{1}{8}$	1·49	1·00	25 × 4	1 × $\frac{3}{16}$	0·952	0·64
30 × 30 × 4	1$\frac{1}{4}$ × 1$\frac{1}{4}$ × $\frac{3}{16}$	2·19	1·47	30 × 4	1$\frac{1}{4}$ × $\frac{3}{16}$	1·190	0·80
40 × 40 × 4	1$\frac{1}{2}$ × 1$\frac{1}{2}$ × $\frac{3}{16}$	2·66	1·79	40 × 3	1$\frac{1}{2}$ × $\frac{1}{8}$	0·952	0·64
50 × 50 × 5	2 × 2 × $\frac{3}{16}$	3·62	2·43	Channels			
60 × 60 × 5	2$\frac{1}{2}$ × 2$\frac{1}{2}$ × $\frac{3}{16}$	4·49	3·02	100 × 50	4 × 2	10·54	7·09
60 × 60 × 6	2$\frac{1}{2}$ × 2$\frac{1}{2}$ × $\frac{1}{4}$	6·01	4·04	150 × 75	6 × 3	18·46	12·41

The brackets and supports carrying the ductwork may have to be fixed to masonry, wood, or steel structures, for which many kinds of fastening are employed. For this purpose, and for connecting the ductwork sections during erection, Whitworth bolts and nuts, studs, and self-tapping screws are used. Heavy square-headed wood screws with washers are used for securing duct brackets to wood beams. The approximate weights of boltings for general ductwork are given in Table 5.

Duct and bracket fastenings are part of the ductwork as installed, and their weight is included in the estimate.

When dealing with warm air systems, ducts fixed in roof spaces, voids, and other unheated spaces must be insulated. These are measured and priced separately.

Table 5. WEIGHTS OF BOLTS, NUTS, WASHERS AND STUDS USED FOR THE ERECTION OF VENTILATING DUCT WORK

Diameter and length from under head		Approx. weight per 100 hexagon Whitworth bolts and nuts		Approx. weight per 100 hexagon Whitworth studs		Approx. weight per 100 circular washers to fit		Approx. weight per 100 self tapping screws and washers Size No. 10 25 mm (1 in) long	
mm	in	kg	lb	kg	lb	kg	lb	kg	lb
6 × 50	$\frac{1}{4}$ × 2	2·09	4·6	1·81	4·0	0·635	1·4	0·454	1·0
10 × 50	$\frac{3}{8}$ × 2	5·62	12·4	3·54	7·8	0·862	1·9		
12·5 × 50	$\frac{1}{2}$ × 2	11·66	25·7	6·80	15·0	1·27	2·8		
15 × 50	$\frac{5}{8}$ × 2	20·77	45·8	11·34	25·0	2·22	4·9		

Note: For a 25 mm (1 in) variation in length of bolts, or studs, add or subtract the following weight per 100:
6 mm ($\frac{1}{4}$ in) dia; 0·635 kg (1·4 lb). 10 mm ($\frac{3}{8}$ in) dia; 1·41 kg (3·1 lb).
12·5 mm ($\frac{1}{2}$ in) dia; 2·49 kg (5·5 lb). 15 mm ($\frac{5}{8}$ in) dia; 3·86 kg (8·5 lb).

INFORMATION REQUIRED BY MANUFACTURER

Fans

When ordering, or requesting a quotation, the following information should be given:

(1) The volume, pressure, speed.
(2) A copy of the main plant arrangement and scheme drawings.
(3) Driving arrangement, i.e. direct drive, or vee-rope drive.
(4) Motor position relative to fan.
(5) Electricity supply available, phase, frequency, volts.
(6) Speed regulation, if required.
(7) Motor type, whether standard, totally enclosed, or flame-proof.
(8) Air operating temperature.
(9) Angle of discharge.
(10) Type of bearings, ball, roller, or ring-oiled sleeve bearings.
(11) Position of lubricators.
(12) Position of inspection door in fan casing, viewed from the driving side of the machine.
(13) Special requirements regarding silence, finishes, etc.
(14) Method of control, manual or thermostatic.
(15) Address of building, or site to which machine is to be delivered.
(16) Latest date for delivery.

Heaters

(1) Inlet air temperature. Outlet air temperature.
(2) Air volume through heater in m³/second.
(3) Type of heater required:

    (*a*) *Electric*. State phase, frequency, volts.
    (*b*) *Hot water*. State temperature.
    (*c*) *Steam*. State gauge pressure.

(4) Method of control, manual or thermostat.
(5) Address of building or site to which heater is to be delivered.
(6) Latest date for delivery.

Air Filters

(1) Air temperature.
(2) Air volume through filter in m³/s.
(3) Any special air condition, i.e. presence of excessive dust, or fumes.
(4) Type of filter required:

    (*a*) Viscous oblique type.
    (*b*) Viscous self-cleaning type.
    (*c*) Cotton wool (throw away) type.
    (*d*) Washable type.
    (*e*) Electrostatic precipitator type.
    (If (*e*) give particulars of electricity supply, phase, frequency, and volts.)

(5) Address of building or site to which the filter is to be delivered.
(6) Latest date for delivery.

Air Washers

(1) Air volume through washer in m³/s.
(2) Air temperature.
(3) Any special air condition, i.e. presence of excessive dust, or fumes.
(4) Detail of proposed plant arrangement.
(5) Position of water pump.
(6) Particulars of water supply.
(7) Particulars of electricity supply, phase, frequency, volts.
(8) Information regarding access to plant room.
(9) Address of building or site.
(10) Latest date for delivery.

Manufacturer's Test and Guarantee

Manufacturers of specialised equipment are prepared to guarantee their products provided they know beforehand the exact amount and nature of the work their equipment will be called upon to do, e.g. a fan maker will be prepared to guarantee a fan to deliver a certain volume of air against a known resistance, at a given speed. Before delivery the machine will be tested to confirm the required performance. Any departure from the operating conditions given to the maker at the time of quotation, e.g. increased resistance due to additional ductwork, or reduced motor speed caused by excess voltage drop in undersized cables, will relieve the maker of his responsibility under the guarantee.

When quoting for expensive items manufacturers are most careful to offer machines or other equipment such as pumps, cylinders, calorifiers, boilers, thermostatic controls, etc., which will adequately do the work specified in the contractor's enquiry.

It is up to the estimator, in the interests of his firm, who are ultimately responsible to the Owner, to provide the maker with comprehensive and accurate information at the time of enquiry. If any doubt exists, even on the most simple point, the estimator should obtain confirmation from the Owner's architect or consulting engineer, before proceeding with the enquiry.

The manufacturer's guarantee regarding quality of materials and workmanship is the Contractor's protection against any claim by the Owner during the guarantee period laid down in the Contract. The estimator will secure the manufacturer's written guarantee at the time the enquiry is made.

Cylinders, boilers, and all vessels which may be subjected to internal working pressure and temperature changes are tested by the maker, from whom a test certificate indicating the maximum safe working pressure of the equipment can be obtained.

The Contractor may be required to supply copies of test certificates to the Owner, and it is better to confirm the manufacturer's liability to provide these when the enquiry is made.

The Contractor is usually held responsible by the Owner for all defects in materials and equipment and for defects in the installation caused by bad workmanship which may arise within a stated period after the satisfactory completion of the Contract.

It is usual to include a clause in the Standard Specification (see Section A, Clause 9, Chapter 2), or in the Terms and Conditions of Contract document, which defines the liability and the maintenance or guarantee period.

Although the responsibility for defects in materials can be passed on to the makers, the full cost of making good defective labour must be borne by the Contractor. The cost of this cannot of course be estimated, but can be reduced to a minimum by good supervision. Serious loss due to shoddy workmanship is a rare occurrence and is not, therefore, a real financial risk on a properly administered contract, with good labour available. However, should it be known at the time of tendering that the employment of mainly new and untried labour will be necessary to complete the work within the contract period, the estimator may cover the risk by adding a percentage to the estimated cost of labour and materials.

The amount to include will depend upon the size of the contract. Generally speaking it is a normal trade risk, but in the circumstances mentioned above, one for which the wise contractor will at least make some provision.

Specials made up in the Contractor's Workshop

Special pipe and other fabrications made in the Contractor's workshop require careful costing. Failure to do this can cause serious financial loss, especially on a large contract where considerable amounts of prefabricated work may be needed.

To make this point clear, suppose a heating return manifold is required for a boiler room.

If ordered from a manufacturer, the Contractor would charge the cost of the manifold as an item in the Bill, to which he adds labour for erection, and a due proportion of overhead charges and profit in order to arrive at his price.

The manufacturer's charges will include cost of materials, his labour for making the manifold, carriage, overhead charges, and profit.

If the Heating Contractor decides to fabricate the manifold in his own workshop the cost should be charged in the Bill of Quantities as an item of materials only and estimated for in the same way as it would if purchased from a manufacturer.

The correct charge to the contract is:

(1) Cost of making manifold, including materials, labour, carriage to site, overheads and profit, which we will assume totals £109·00.

(2) The estimate for tender will be: £
    (*a*) Cost of materials (No. 1 manifold)   109·00
    (*b*) Cost of labour for erection, National
        Insurance, etc.   16·25

Carried forward   125·25

Brought forward   125·25   125·25
(c) Overhead charges, and profit, say 15 per cent   18·78

144·03

Amount to be included in tender   £144·00

The point to note is that the Contractor must cost the making of the item in the same way as the manufacturer, and enter it in the Bill as a purchased item, which in fact it is, because from a costing point of view it has been purchased from the Contractor's workshop. If the workshop is to pay its way any work undertaken there should be separately costed and charged to the particular contract.

Prefabrication of pipework sections and other assemblies for all parts of the installation is having an increasing effect in improving the productivity of the industry.

Work which can be prefabricated under workshop conditions is cheaper than similar work done on site, particularly during the rough early stages of building construction, and during the winter period when discomfort slows down production on the site.

Provided a constant flow of work is assured, an adequately tooled and staffed workshop constitutes a good investment.

HAULAGE OF MATERIALS

Materials, when ordered for a specific contract, are usually delivered to the site by the manufacturer with his own vehicle, or by public transport. The price quoted by the supplier will include free delivery in given circumstances, and in any case it is better to obtain quotations for materials *delivered to the job*.

For the haulage of materials within a radius of 25 to 30 miles from his stores, the contractor will make use of his own petrol-driven van or lorry.

Haulage costs may be divided into running costs and standing charges. Running costs, which depend upon mileage, are made up of petrol and oil, repairs, tyres, and maintenance. Standing charges include driver's wages, insurance, rent of garage, Road Fund Tax, and depreciation.

The amount to charge in a particular tender depends upon the estimated number of hours the driver and vehicle will be needed during the period of the contract, and the cost, per mile or per hour, of the vehicle and driver.

For heating estimates cost per hour is a more useful factor than cost per mile. Owing to standing time during unloading, and other delays at the stores or on the site, and delay due to traffic congestion, the time taken to complete two journeys of exactly equal mileage may be quite different.

The items which make up the cost of road haulage may of course change from day to day. Petrol, oil, driver's wages, maintenance and repairs, are variables which prohibit a really close estimate of future costs.

Based upon present day prices the operating costs of a heating contractor's 1·5–2·0 tonne (30 cwt) petrol-driven lorry can be estimated as shown in the following example.

For the purpose of the estimate it is assumed that the lorry will average three round journeys per working day of 8 hours, each a distance of 20 miles, or alternatively, two round trips of 30 miles. On this basis the weekly mileage will be:

5 days × 60 miles = 300 miles.

The annual mileage will be 52 weeks × 300 miles = 15 600 miles.

Due to waiting time between journeys, and periods when the lorry is laid up for repairs, it is estimated that a Heating Contractor's lorry is gainfully employed for not more than 75 per cent of its full time. The full working time of the vehicle is related to the 40 hour working week of the driver.

The annual working time will be 52 weeks × 40 hours = 2080 hours.

Operating Costs for 1·5–2·0 tonne Lorry based on an Estimated Annual Mileage of 24 960 km (15 600 miles)*

Cost of new lorry = £1 100.
Life of vehicle, say 7 years.

Annual standing charges: £

Depreciation = $\dfrac{£1100}{7 \text{ years}}$ =   157·14

Driver's wages at £16 per week   832·00
Employer's National Insurance and Redundancy Fund
　contribution, at £0·95 per week × 52   49·40
Holidays with Pay, Sickness and Accident Premiums at
　mate's rate of £1·185 per week × 52 weeks   61·62
S.E.T., £1·20 × 52 weeks   62·40
Garage rent at £1·00 per week   52·00
                                                                       ———
                                               Carried forward   1214·56

	Brought forward	1214·56
Insurance of vehicle		50·40
Road Fund licence		63·00
	Total standing charges	1327·96

Annual running costs: £

Petrol, 24 960 km at 7·036 km/litre, 3550 litres at 7p per litre	248·50
Oil for engine lubrication, estimated at 354 km/litre. 69 litre at 26p/litre	18·34
Oil and grease for gearbox and back axle	4·00
One new set of tyres	65·70
Add 15 per cent of cost of vehicle for repairs and maintenance, spare parts, washing, cleaning, and repainting. 15 per cent of £1 100	165·00
Total annual costs	501·54
Total annual standing charges	1327·96
Total annual operating costs	1829·50
Say	£1 830·00

As the lorry will be gainfully employed for only three-quarters of the annual working time, the time in hours upon which to base the cost per hour is:

$$\frac{2\,080 \text{ hours} \times 3}{4} = 1560 \text{ hours.}$$

The vehicle cost per hour for a 1·5–2·0 tonne petrol-driven lorry on heating contract work will be:

$$\frac{\text{Total annual operating costs}}{\text{Annual hours lorry is gainfully employed}} = \frac{£1830}{1\,560 \text{ hours}}$$
$$= £1·175 \text{ per hour.}$$

* For conversion factors: miles to km, and km to miles, see Appendix B.

This figure of £1·175 is the net cost to the heating contractor, and is the usual hourly charge to include in the Bill for the transport of the Contractor's own materials to or from a contract site. As with other material and labour items in the Bill, it will then be subject to the addition of overhead charges and profit.

When employed on jobbing work, the rate for costing will be £1·175 per hour, plus overheads and profit.

Suppose the haulage time for a particular job, as taken by the accounts clerk from the lorry driver's time sheet, is 10 hours. The item will be costed as follows:

	£
Lorry, and driver, 10 hours at £1·175	11·75
Overheads, and profit, at 20 per cent	2·35
Total charge	14·10

which works out at approximately £1·41 per hour.

For a 0·75 tonne truck, the charge, based on present-day costs, and including 20 per cent for overheads and profit, is about £1·20 per hour. As covering figures when preparing estimates, the following can be used:

Rate for 1·5 tonne lorry with driver. £1·40 per hour.
Rate for 0·75 tonne truck with driver. £1·20 per hour.

The contractor may wish to transport men to and from jobs within the 50 km (30 miles) radius, in lieu of payment of return daily travelling fares.

Up to five passengers may travel in a suitable 1·5 tonne goods vehicle, provided the insurance cover includes for passenger liability. This risk can be covered by the payment of an increased premium, which usually amounts to not more than 5 per cent of the normal comprehensive policy premium.

The insurance item in the foregoing make-up of annual standing charges includes for full comprehensive cover plus passenger liability.

THE CARE OF MATERIALS ON SITE

The Standard Specification invariably includes a clause (see Section A, Clause 8) defining the Heating Contractor's responsibility for the reception and storage of materials at the site.

Proper storage is advantageous both to the Owner and the Contractor, especially when large quantities of valuable tubes, fittings, valves, and other equipment have to be kept on the site.

Tube should be stored in metal or wooden racks and sheeted over. Fittings and valves, sorted into their various types and sizes, are best kept in a locked site hut, to be issued by the charge-hand or the job foreman as needed.

Ventilating ductwork, radiators, and boiler sections can be laid across planks off the ground, and securely sheeted.

The cost of proper storage is a good investment, as it not only limits pilfering but ensures pipes and fittings remaining in good condition until needed for erection, and also eliminates the costly possibility of those blocked circuits which are so often discovered when the installation is tested.

Pipes, radiators, and boiler sections left lying unprotected around the site attract to their 'innards' mud, dust, and all kinds of foreign matter, which sooner or later cause trouble in the system. If the trouble occurs within the Contract Guarantee Period, usually 12 months, the Heating Contractor may have to spend weeks of labour dismantling, clearing, and re-erecting whole sections of pipework, radiators, and other equipment. This represents a serious financial loss to any Contractor, quite apart from loss of goodwill. When such breakdowns occur, as they often do, after the guarantee has expired, the responsibility for repairs falls upon the Owner, who, together with his architect or consulting engineer, must in their own interests try to avoid a repetition of these troubles when considering invitations to tender for future work.

4

ESTIMATING LABOUR REQUIRED FOR ERECTION

Estimating labour is not, and never can become, an exact science; the knowledge and experience of the skilled estimator are therefore essential to successful tendering.

Of the various items which make up the tender, labour is the most difficult because it is subject to variable features, the effects of which cannot be predicted. The weather, site conditions, pace of job, quality of labour, breaks in labour and supervision continuity due to sickness, holidays, or the labour demands of other jobs, are a few of the uncontrollable factors which can upset the most careful estimate.

The estimation of time required for erection is therefore a matter of experience. Most offices possess schedules of labour rates, based upon information accumulated during years of trading, which provide a reliable basis for estimating. While this information is valuable to the firm concerned it is unlikely that the labour time rates used by any two firms will be completely identical.

The tables which follow have been compiled by the author during many years in the industry and are offered as a useful guide for estimating labour costs.

The labour constants given in the tables represent the time in hours taken by one fitter and one adult mate to complete each process. Certain jobs, such as erecting large boilers, cylinders, or other heavy equipment which require more than one pair of men, will of course be completed in less time than stated in the tables. For estimating purposes, however, the basic labour unit remains the same, viz: 1 fitter and 1 mate, because although the increased manpower reduces the time taken to complete the work, the total labour cost remains the same.

In the following pages tables of labour constants are arranged in sections according to the type of materials or equipment to be installed.

To facilitate the use of the tables an index is included.

INDEX TO LABOUR ESTIMATING TABLES

Section 1

STEEL PIPES AND FITTINGS

Off load at site, pipe, sockets, fittings, valves, and move to fixing position on ground floor of new building. Arrange pipe run from plan, level and mark out for brackets, cut pipe to required lengths, screw for and fit sockets. Fix pipework to brackets at low level on wall, and clean off joints.

Measure up for valves and fittings, cut and screw two threads per fitting (one thread per flange). Fit valve or other fitting into pipe line and clean off joints.

A. Pipe per metre run.

B. Screwed valves and fittings.

C. Flanged valves and fittings.

D. Fit one flange. Thread and expand.

Nominal pipe size (bore)		Hours (fitter and mate)			
mm	in	A	B	C	D
15	½	0·15	0·35		
20	¾	0·18	0·40		
25	1	0·19	0·50		
32	1¼	0·20	0·55		
40	1½	0·24	0·65		
50	2	0·30	0·80		
65	2½	0·35	0·90	1·50	0·60
80	3	0·42	1·00	1·50	0·60
100	4	0·60	1·50	2·25	1·00
125	5	0·75	2·00	2·75	1·20
150	6	1·00	2·50	3·25	1·50

Variations

For work in occupied buildings add 20 per cent.
For work on scaffold or ladder above 3 m and up to 6 m add 10 per cent.
For work on scaffold or ladder at heights between 6 m and 9 m add 15 per cent.
For work in restricted subways, trenches add 30 per cent.
For work above or below ground floor add 5 per cent for each floor.

Note: For pipe run in prepared earth trench on open ground allow 40 per cent of times in column 2.

Example. Estimate the time required to erect 30 m of 40 mm steel pipe, 4 bends, 8 tees and 2 stop valves, on the second floor of a new building at a height of 4·60 m above floor level.

From the table:		*hours*
30 m × 0·24	=	7·2
4 bends × 0·65	=	2·6
8 tees × 0·65	=	5·2
2 valves × 0·65	=	1·3
	Carried forward	16·3

	Brought forward	16·3
Add 10 per cent for second floor		1·6
Add 10 per cent for working at high level		1·6
Total estimated time		19·5

Section 2

LIGHT GAUGE COPPER PIPE AND FITTINGS
(B.S. 2871: Part 1)

Off load pipe, sockets, and brackets at site, and move to fixing position on ground floor of new building. Arrange pipe run from plan, level and mark out for brackets, cut pipe to required length and fit sockets (soldered or compression joints). Fix pipework to brackets at low level on wall, and clean up joints. Off load fittings and valves at site and move to fixing position. Measure up for fixing, cut pipe and fit compression or soldered capillary fitting into the line and clean up joints.

Nominal pipe size		Pipe per m run		Tee, bend, valve, etc.	
B.S.2871 mm (o.d.)	B.S.659 in (i.d.)	Capillary	Compression	Capillary	Compression
12	$\frac{3}{8}$	0·10	0·09	0·25	0·20
15	$\frac{1}{2}$	0·11	0·09	0·30	0·25
22	$\frac{3}{4}$	0·13	0·10	0·30	0·25
28	1	0·15	0·13	0·35	0·30
35	$1\frac{1}{4}$	0·18	0·15	0·40	0·35
42	$1\frac{1}{2}$	0·20	0·19	0·45	0·40
54	2	0·24	0·22	0·60	0·50
76·1	3	0·33	0·28	1·00	0·70
108	4	0·42	0·38	1·50	0·90

Note: Variation percentages in Section 1 apply also in Section 2.

Section 3

BENDING STEEL OR COPPER PIPES

Steel pipe

Off load pipe at site and move to fixing position on ground floor of new building. Mark off pipe length and cut. Mark off pipe for bending, bend pipe, cut and screw two threads, fit bend or offset into pipe line and clean off joints. Bends and offsets 65 mm (2½ in) bore and above will be fitted with 2 flanges and 2 counter flanges for fixing.

Copper pipe

Off load at site and move to fixing position on ground floor of new building. Mark off pipe length and cut. Mark off pipe for bending, anneal and bend pipe, fit bend or offset into pipe line and clean off joints.

SECTION 3. *(Contd.)*

				Hours, fitter and mate				
	Steel pipe, *B.S.1387*					Copper pipe, *B.S.2871*		
Nominal pipe size (bore)		*Fire bend*		*Machine bend*		*Pipe size*		*Using spring or machine*
mm	*in*	*Single bend*	*Offset*	*Single bend*	*Offset*	*mm (o.d.)*	*in (i.d.)*	*Single bend*	*Offset*
15	$\frac{1}{2}$	1·00	1·50	0·90	1·25	15	$\frac{1}{2}$	0·80	1·50
20	$\frac{3}{4}$	1·25	1·65	1·10	1·40	22	$\frac{3}{4}$	0·80	1·50
25	1	1·50	2·00	1·25	1·60	28	1	1·00	1·90
32	$1\frac{1}{4}$	1·75	2·25	1·50	1·80	35	$1\frac{1}{4}$	1·25	2·25
40	$1\frac{1}{2}$	2·00	2·50	1·75	2·10	42	$1\frac{1}{2}$	1·50	2·80
50	2	2·25	3·00	2·00	2·50	54	2	2·00	3·60
65	$2\frac{1}{2}$	3·00	4·00	2·75	3·50				
80	3	4·00	5·00	3·60	4·75				
100	4	5·00	6·00	4·60	5·30				

Variations
For work in occupied buildings add 20 per cent.
For work above or below ground floor add 5 per cent for each floor.

Section 4(A)
WELDING STEEL PIPEWORK

Off load pipe at site, and move to fixing position on ground floor of new building.
Move equipment to fixing position. Prepare and mark out, cut and shape hole in
main pipe, cut and shape branch pipe, set up job and weld.

Nominal pipe size (bore)		*h = Hours, fitter-welder and mate*							
mm	*in*	*Butt weld pipes* h	*Acet. gas* m^3	*Oxy. gas* m^3	*Welding wire* kg	*Bend or plain tee* h	*Acet. gas* m^3	*Oxy. gas* m^3	*Welding wire* kg
15	$\frac{1}{2}$	0·25	0·08	0·08	0·09	0·60	0·11	0·14	0·09
20	$\frac{3}{4}$	0·25	0·11	0·14	0·14	0·65	0·14	0·17	0·14
25	1	0·30	0·14	0·17	0·18	0·70	0·17	0·20	0·18
32	$1\frac{1}{4}$	0·35	0·17	0·20	0·20	0·80	0·20	0·23	0·20
40	$1\frac{1}{2}$	0·40	0·20	0·24	0·23	0·90	0·23	0·28	0·23
50	2	0·50	0·23	0·25	0·27	1·25	0·25	0·31	0·32
65	$2\frac{1}{2}$	0·60	0·23	0·28	0·29	1·50	0·28	0·37	0·36
80	3	0·70	0·25	0·31	0·32	2·00	0·34	0·40	0·41
100	4	1·00	0·34	0·45	0·41	2·50	0·40	0·51	0·45
125	5	1·20	0·40	0·54	0·54	3·00	0·45	0·62	0·59
150	6	1·50	0·48	0·62	0·68	3·50	0·51	0·65	0·73

Variations (on labour only)
For work in occupied buildings add 20 per cent.
For work in cramped conditions in ducts, trenches, and subways add 30 per cent.
For work at high level in room add 20 per cent.
For each floor above or below ground floor add 5 per cent.

Section 4(B)
WELDING STEEL PIPEWORK

Off load pipe and fittings at site and move to fixing position on ground floor of new building. Move equipment to fixing position. Prepare and mark out. Cut and shape, set up job and weld.

Nominal pipe size (bore)		h = Hours, fitter-welder and mate							
		Pitcher tee	Acet. gas	Oxy. gas	Welding wire	Flange	Acet. gas	Oxy. gas	Welding wire
mm	in	h	m³	m³	kg	h	m³	m³	kg
15	½	0·70	0·14	0·17	0·11	0·30	0·09	0·11	0·11
20	¾	0·75	0·17	0·20	0·16	0·35	0·13	0·16	0·16
25	1	0·80	0·20	0·25	0·20	0·40	0·14	0·17	0·20
32	1¼	1·00	0·27	0·34	0·23	0·50	0·17	0·20	0·25
40	1½	1·25	0·31	0·38	0·32	0·70	0·21	0·25	0·32
50	2	1·50	0·33	0·40	0·41	0·80	0·28	0·35	0·41
65	2½	2·00	0·38	0·50	0·50	0·90	0·31	0·40	0·43
80	3	2·50	0·40	0·55	0·57	1·00	0·37	0·45	0·45
100	4	3·00	0·51	0·65	0·68	1·25	0·48	0·62	0·57
125	5	3·50	0·57	0·78	0·79	1·40	0·57	0·76	0·79
150	6	4·00	0·76	0·91	0·91	1·60	0·65	0·82	0·91

Section 5
MAKING AND FITTING STEEL PIPE SLEEVES ON THE SITE

A. Off load pipe at site and move to bench on ground floor of new building. Mark off, cut and screw two threads, and fit Hall's thimbles. Fit completed sleeve to pipe, set into position in wall or floor ready for the builder to make good.
B. As above, but for plain unscrewed sleeves with wall or floor plates.

Size of pipe, for sleeve		To pass pipe size		Hours, fitter and mate	
mm	in	mm	in	A	B
25	1	15	½	1·00	0·50
32	1¼	20	¾	1·10	0·55
40	1½	25	1	1·25	0·60
50	2	32	1¼	1·40	0·65
65	2½	40	1½	1·50	0·75
80	3	50	2	1·75	0·80
100	4	65	2½	2·50	0·90
100	4	80	3	2·50	0·90
125	5	100	4	3·00	1·00

Variations
For work in occupied buildings add 20 per cent.
For work on each floor above or below ground floor add 5 per cent.

Section 6
CAST IRON PIPES

A. Off load at site, pipes, brackets, and jointing materials, and move to fixing position on ground floor of new building. Set out pipe run from plan, mark out for brackets, lay pipes on brackets, and level up ready for jointing.

B. Caulk one joint on straight length, and clean up.

C. Make one rubber joint on straight length, and clean up.

D. Off load at site, valve, fitting, jointing materials, mark off and cut pipe for valve or fitting, caulk two joints, and clean off.

E. As **D** for rubber-jointed valves and fittings.

F. Off load at site, pipe, spigot ring, portable forge, and fuel. Mark off and cut pipe, and fit iron spigot ring.

Nominal pipe size (bore)		*Hours, fitter and mate*					
mm	in	**A** per metre run	**B**	**C**	**D**	**E**	**F**
51	2	0·18	0·3	0·20	1·00	0·60	1·00
76	3	0·24	0·5	0·35	1·50	0·90	1·50
102	4	0·30	0·7	0·50	2·00	1·25	2·00

Variations

For work in occupied buildings add 20 per cent.

For work at high level in room add 20 per cent.

For work in restricted ducts, trenches, and subways add 30 per cent.

For work on each floor above or below ground floor add 5 per cent.

Section 7
CAST IRON PIPEWORK

Off load pipe and saddle at site and move to fixing position on ground floor of new building. Mark off and drill main, and fit one tapping pipe saddle to form branch.

Branch size		Hours, fitter and mate
mm	in	
20	$\frac{3}{4}$	1·0
25	1	1·5
32	$1\frac{1}{4}$	2·0
40	$1\frac{1}{2}$	2·5

Note: Variation percentages in Section 6 apply also to Section 7.

Section 8

RADIATORS (CAST IRON AND STEEL)
(See Figure 7)

Off load at site radiators, valves, fittings, and pipe for radiator connections, and move to fixing position on ground floor of new building. Mark out for radiator brackets and top stay. Mount radiator on brackets after building-in by bricklayer. Cut and screw main and fit two tees for radiator connections. Measure up for, cut and screw pipe for radiator connections, fit valves and fittings, completing connection of radiator to mains. Clean all joints, and take down and refix once for painter.

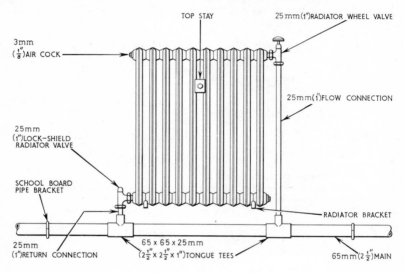

Figure 7

Three principal factors vary the time required to install a radiator or other form of heater connected to a hot water or steam system. These are:

(1) Size of radiator flow and return connections.
(2) Amount of work needed to connect radiator flow and return to mains.
(3) Weight of radiator.

The time constants given in Section 8 and in the other sections dealing with space heaters have therefore been related to these factors rather than to thermal output.

When estimating time by the use of these labour constants it should be remembered that the results obtained include for off-loading, carrying of materials, tools, and where necessary setting

Size of main		Hours, fitter and mate					
		Cast iron radiators Connections			Steel radiators Connections		
mm	in	15–20 mm $\frac{1}{2}-\frac{3}{4}$ mm	25 mm 1 in	32 mm $1\frac{1}{4}$ in	15–20 mm $\frac{1}{2}-\frac{3}{4}$ mm	25 mm 1 in	32 mm $1\frac{1}{4}$ in
25	1	4·50			3·75		
32	$1\frac{1}{4}$	4·75	5·25		4·00	4·50	
40	$1\frac{1}{2}$	5·00	5·50	6·00	4·25	4·75	5·50
50	2	5·25	5·75	6·25	4·75	5·25	5·75
65	$2\frac{1}{2}$	5·50	6·00	6·50	5·00	5·50	6·00

Variations
For fitting radiator to existing circuit add 30 per cent.
For work in occupied building add 30 per cent.
For connecting to mains in trench, floor space, or subway add 30 per cent.
For connecting to mains at high level on floor below add 20 per cent.
For each floor above or below ground floor add 5 per cent.

Note: Where radiators have to be taken down more than once, extra time should be allowed. Include 0·5 h for each additional taking down and refilling operation.

up scaffolds and ladders at fixing positions, in fact all processes from the time the materials arrive at site, in addition to the actual work of erection.

Example. Estimate the labour required to erect the pipework and radiators shown in Figure 8.

(Reference Sections 1, 5, and 8.)

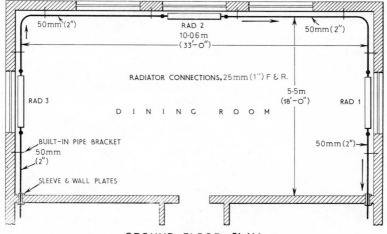

GROUND FLOOR PLAN
Figure 8

The pipe is first measured from the drawing, followed by the wall sleeves and, last, the radiators.

1	2	3	4
No.	Size or description	Labour constant	Estimated time in hours Col. 1 × Col. 3
21·04	metres of 50 mm pipe	0·30 hours Section 1	6·31
2	50 mm bends	0·80 hours Section 1	1·60
2	wall sleeves for 50 mm pipe with Hall's Thimbles	1·75 hours Section 5	3·50
3	cast iron radiators. 25 mm connections from 50 mm main	5·75 hours Section 8	17·25

Total time = 28·66 hours

Section 9

SKIRTING HEATING PANELS

Off load panel sections, accessories, pipe, fittings, and valves for connections at site, and move to fixing position on ground floor of new building. Assemble heater, fix brackets, fit sleeves and floor plates, fit valves, steam traps, strainers, and connect up. Clean all joints. (Mains in trenches or at high level on floor below.)

Water-heated panels

Size of main		Hours, fitter and mate		
		Size of connections		
mm	in	15–20 mm $\frac{1}{2}$–$\frac{3}{4}$ in	25 mm 1 in	32 mm $1\frac{1}{4}$ in
25	1	5·50		
32	$1\frac{1}{4}$	6·00	6·50	7·00
40	$1\frac{1}{2}$	6·25	6·75	7·25
50	2	6·75	7·50	8·00

Steam-heated panels
Steam at 0·35 to 0·7 bar (5 to 10 lb/in²)

Size of main		Hours, fitter and mate	
		Size of connections	
mm	in	15 mm $\frac{1}{2}$ in	20 mm $\frac{3}{4}$ in
20	$\frac{3}{4}$	6·00	6·25
25	1	6·25	6·50
32	$1\frac{1}{4}$	6·50	6·75
40	$1\frac{1}{2}$	6·75	7·00

Variation
For work on each floor above or below ground add 5 per cent.

Section 10

CONVECTORS (WATER AND STEAM)

Off load convectors, valves, fittings, and pipe for connections, at site. Move to fixing position on ground floor of new building. Mark off for fixing. Dismantle convector casing and fix heater element on brackets.

Cut and screw main and fit tees. Cut and screw connections, fit sleeves and floor plates, fit valves, steam traps, and strainers and connect up. Refit convector casing. Clean all joints.

Size of main		Hours, fitter and mate			
		Size of connections			
		Water		Steam, 0·35 to 0·70 bar (5 to 10 lb/in²)	
mm	in	15–20 mm $\frac{1}{2}-\frac{3}{4}$ in	25 mm 1 in	15 mm $\frac{1}{2}$ in	20 mm $\frac{3}{4}$ in
25	1	6·00		6·50	6·75
32	$1\frac{1}{4}$	6·25	6·75	6·75	7·00
40	$1\frac{1}{2}$	6·50	7·00	7·00	7·25
50	2	6·75	7·25		
65	$2\frac{1}{2}$	7·50	8·00		
80	3	7·75	8·25		
100	4	8·75	9·25		
125	5	9·75	10·25		
150	6	10·25	10·75		

Variations
For fitting convector to existing circuit add 30 per cent.
For work in occupied building add 40 per cent.
For each floor above or below ground floor add 5 per cent.

Section 11

CONCEALED HOT WATER CEILING PANELS
15 mm ($\frac{1}{2}$ in) pipes at 150 mm (6 in) centres (see Figure 9)

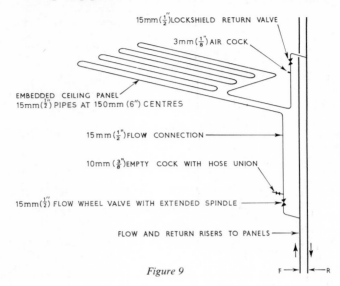

Figure 9

A. Off load at site, panels, pipe, fittings, and valves, and move to fixing position. Lay panel on first floor shuttering, level and secure. Fit or weld two tees in risers, and fit valves, empty cock and air cock. Remove end caps, joint tails and connect panel to risers as shown in Figure 9. Test panels as specified and fit valve plates to wall.

B. As above, but panel fixed in suspended ceiling. Wire panel to fixing bars fixed by builder.

Size of rising mains		Hours, fitter-welder and mate	
mm	in	A	B
20	$\frac{3}{4}$	7·00	8·50
25	1	7·20	8·70
32	$1\frac{1}{4}$	7·30	8·80
40	$1\frac{1}{2}$	7·50	9·00
50	2	7·75	9·25
65	$2\frac{1}{2}$	8·00	9·50

Variation
For work on each floor above first floor add 5 per cent.

Section 12
HOT WATER FLOOR PANELS
20 mm ($\frac{3}{4}$ in) pipes at 190 mm ($7\frac{1}{2}$ in) or 225 mm (9 in) centres

Off load at site, panels, pipes, fittings, and valves, and move to fixing position on ground floor. Lay on floor slab, level, remove end caps, fit tees in mains, valves on flow and return connections, connect to mains in adjacent trench, fill up and test as specified.

Size of mains		Hours, fitter-welder and mate
mm	in	
25	1	6·00
32	$1\frac{1}{4}$	6·20
40	$1\frac{1}{2}$	6·50
50	2	6·75
65	$2\frac{1}{2}$	7·00
80	3	7·50
100	4	8·00

Section 13

STEAM HEATED RADIANT PANELS FIXED ABOVE FLOOR IN PAIRS AS FIGURE 10
Empty weight of each panel 77 kg (170 lb) to 91 kg (200 lb)

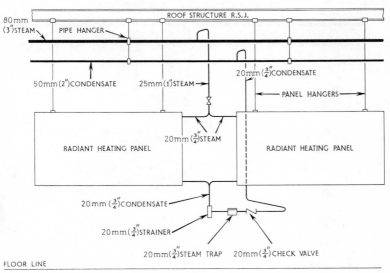

Figure 10

Off load at site, panels, pipes, valves, fittings, hangers, steam traps, strainer, and move to fixing position. Arrange scaffolding for fixing. Hoist up and hang panels. Measure up for connecting pipework. Fit tees in steam and condense mains. Fit valve, steam trap, check valve and strainer, and connect panels to mains, as shown in Figure 10.

Size of steam main		Hours per pair fitter and mate
mm	in	
32	1¼	14·00
40	1½	14·00
50	2	14·50
65	2½	15·00
80	3	15·50
100	4	16·00
125	5	17·00
150	6	18·00

Section 14

RADIANT PANELS (HIGH TEMPERATURE HOT WATER) IN PAIRS AS FIGURE 11
(Empty weight of each panel 77 kg (170 lb) to 91 kg (200 lb))

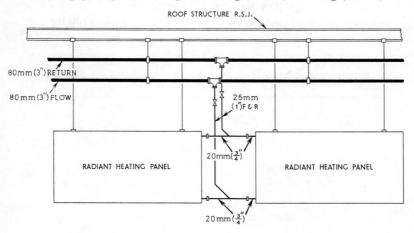

Figure 11

Off load at site, panels, pipes, valves, fittings, hangers, and welding equipment, and move to fixing position. Arrange scaffolding for fixing. Fix hangers and brackets. Hoist up and hang panels. Measure up for connecting pipework. Weld flanges in connecting pipework for valves and fittings, and connect to mains and panels, as shown in Figure 11.

| Size of mains | | Hours per pair |
mm	in	fitter-welder and mate
32	1¼	17·00
40	1¼	17·50
50	2	18·00
65	2½	18·50
80	3	19·00
100	4	20·00
125	5	20·50
150	6	21·50

Section 15

STEAM UNIT HEATERS FIXED AT HIGH LEVEL, AND CONNECTED AS SHOWN IN FIGURE 12

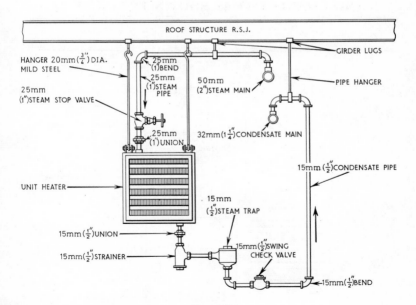

Figure 12

Off load at site, heater, piping fittings, valves, steam trap, check valve, strainer, hangers, brackets, and supports, and move to fixing position at ground level. Set up scaffold or fixing platform. Measure up. Cut and screw mains and fit tees. Fix supports, brackets, or hangers. Hoist up and hang heater. Cut and screw connecting pipework and connect heater to steam and condense mains as shown in Figure 12.

In the following table: **S** = Steam pipe   **C** = Condensate pipe

Heater rating (up to)		Approx. weight	Size of connections		Hours, fitter and mate							
					Size of mains, mm							
			mm		**S** 32	40	50	65	80	100	125	150
kW	Btu/h	kg	**S**	**C**	**C** 25	32	32	40	50	65	80	80
14·66	50 000	39	25	25	12·0	13·0	13·5	14·0	14·5	15·0	15·5	16·0
23·45	80 000	45	32	20		15·0	15·5	16·0	16·5	17·0	17·5	18·0
35·17	120 000	57	32	20		17·0	17·5	18·0	18·5	19·0	19·5	20·0
49·83	170 000	73	40	25			20·0	21·0	21·5	22·0	23·0	24·0
70·34	240 000	86	40	25			21·5	22·0	22·5	23·0	24·0	25·0
87·93	300 000	100	40	32			22·5	23·0	23·5	24·0	25·0	26·0

Section 16
HOT WATER UNIT HEATERS FIXED AT HIGH LEVEL, AND CONNECTED AS SHOWN IN FIGURE 13

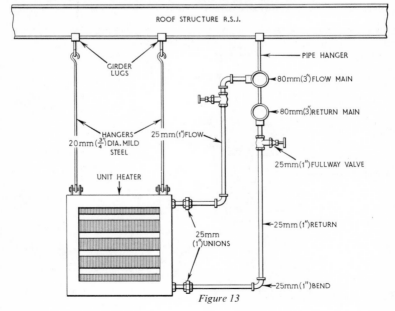

Figure 13

Off load at site, heater, piping, fittings, valves, hangers, brackets, and supports, and move to fixing position at ground level. Set up scaffold or fixing platform. Fix hangers and brackets. Measure up for connections. Cut and screw mains and fit tees. Hoist up and hang heater. Cut and screw connecting pipework, and connect heater to flow and return mains as shown in Figure 13.

Heater rating		Approx. weight	Size of connections	Hours, fitter and mate							
				Size of mains, mm							
kW	Btu/h	kg	mm	32	40	50	65	80	100	125	150
14·66	50 000	39	25	8·0	9·0	10·0	10·5	11·0	11·5	12·5	13·0
23·45	80 000	45	25	9·0	10·0	11·0	11·5	12·0	12·5	13·5	14·0
35·17	120 000	57	25	10·0	11·0	12·0	12·5	13·0	13·5	14·5	15·0
49·83	170 000	73	32		12·5	13·0	13·5	14·0	14·5	15·0	16·0
70·34	240 000	86	32		13·5	14·0	14·5	15·0	15·5	16·0	16·5
87·93	300 000	100	32		14·5	15·0	15·5	16·0	16·5	17·0	17·5

Section 17

HEATING AND HOT WATER SUPPLY BOILERS

A. Erect in ground floor boiler room.
B. Erect in basement boiler room.
C. Fit steel insulating jacket.

Off load at site, boiler or boiler sections, fire bars, fire doors, hand damper gear, smoke hood, safety valve, one empty cock, altitude gauge and thermometer, and move to fixing position. Mark out boiler base for builder, erect boiler on base, fit fire doors, smoke hood, hand-operated damper gear and full set of mountings mentioned above. (Connection to chimney not included. For additional time see Section 19.)

Boiler heating surface		Hours, fitter and mate		
m²	ft²	A	B	C
1·39	15	8	9	1·5
1·86	20	10	11	1·5
2·79	30	14	16	1·5
3·72	40	16	18	2·0
4·65	50	20	22	2·5
9·29	100	23	26	3·0
13·94	150	25	31	3·0
18·58	200	40	47	3·0
23·23	250	45	54	3·5
27·87	300	53	61	4·0
37·16	400	73	81	4·0
46·45	500	100	110	5·0

Note: For work in boiler room of occupied building with unrestricted space and good access add 10 per cent.

Section 18

FITTING BOILER MOUNTINGS WHEN FIXED AS SEPARATE ITEMS

Off load at site, move to boiler room, and fit to boiler.

Boiler heating surface		Hours, fitter and mate					
m^2	ft^2	Safety valve and drip pipe	Thermometer	Altitude gauge	Thermostat	Front damper automatic control	Front and check dampers motor control
1·39	15	1·0	0·5	0·5	0·5	1·0	5·5
1·86	20	1·0	0·5	0·5	0·5	1·0	5·5
2·79	30	1·0	0·5	0·5	0·5	1·0	5·5
3·72	40	1·0	0·5	0·5	0·5	1·0	5·5
4·65	50	1·0	0·5	0·5	0·5	1·0	5·5
9·29	100	1·5	0·5	0·5	0·8	1·5	6·0
13·94	150	1·5	0·5	0·5	0·8	1·5	6·0
18·58	200	1·5	0·5	0·5	0·8	1·5	6·0
23·23	250	1·5	0·5	0·5	0·8	1·5	6·0
27·87	300	2·0	0·8	0·8	0·8	1·5	6·5
37·16	400	2·0	0·8	0·8	0·8	1·5	6·5
46·45	500	2·0	0·8	0·8	0·8	1·5	6·5

Section 19

SMOKE PIPE AND FITTINGS

Off load at site, smoke pipe and fittings, move to boiler room and connect boiler to brick chimney. Fit brackets and stays and pack between smoke pipe and metal sleeve in chimney with asbestos rope.

Diameter of smoke pipe		Hours, fitter and mate				
mm	in	Straight 1·0 m or 1·3 m (3 ft or 4 ft) connections	As Col. 2 with one bend	As Col. 2 with two bends	One bend with 1·83 m (6 ft) straight connection	Two bends with 1·83 m (6 ft) straight connection
100	4	1·0	1·5	2·0	2·0	2·5
115	4·5	1·0	1·5	2·0	2·0	2·5
125	5	1·0	1·5	2·0	2·0	2·5
150	6	1·0	1·5	2·0	2·0	2·5
175	7	1·5	2·0	2·5	2·5	3·0

Diameter of smoke pipe		Hours, fitter and mate				
		Straight 1·0 m or 1·3 m (3 ft or 4 ft) connections	As Col. 2 with one bend	As Col. 2 with two bends	One bend with 1·83 m (6 ft) straight connection	Two bends with 1·83 m (6 ft) straight connection
mm	*ln*					
200	8	1·5	2·0	2·5	2·5	3·0
225	9	1·5	3·0	3·5	3·5	4·0
250	10	1·5	3·0	3·5	3·5	4·0
300	12	2·0	4·0	4·5	4·5	5·0
350	14	2·5	4·5	5·0	5·0	5·5
400	16	3·5	5·0	6·0	6·0	6·5
450	18	5·0	6·5	7·5	7·5	8·5
500	20	6·5	8·0	9·0	9·0	10·0

Section 20

HOT WATER STORAGE CYLINDERS

Off load at site, one cylinder, altitude guage, thermometer, empty cock, and brackets, or other steel supports. Move to fixing position in plant room of new building. Mark out bracket fixings or masonry foundation slab for builder. Set cylinder on brackets or slab, level up, fit mountings, and clean off joints, all ready for piping connections.

A. Ground floor plant room.   **B.** Basement plant room.

Cylinder capacity litres	Cylinder capacity gallons	Hours, fitter and mate			
		Direct cylinder		Indirect cylinder	
		A	**B**	**A**	**B**
182	40	5·0	5·5	6·0	7·0
273	60	6·0	6·5	7·0	8·0
364	80	7·0	7·5	8·0	9·0
455	100	8·0	9·0	9·5	10·5
680	150	9·5	10·5	11·0	12·0
910	200	11·0	12·0	12·5	13·5
1 365	300	13·0	14·5	15·0	16·0
1 820	400	17·0	19·0	19·0	21·0
2 275	500	22·0	24·0	25·0	28·0

Variations
For work in plant room of occupied building with unrestricted space and good access add 10 per cent.
For fixing cylinders in other rooms above ground floor, for each floor add 10 per cent.
For work above ground floor of occupied building add 30 per cent.

Section 21

STEAM CALORIFIERS

Off load at site, one steam calorifier, altitude gauge, air release valve, safety valve, thermometer, steam trap, check valve, empty cock, and steel brackets or supports. Move to fixing position in plant room of new building. Mark out bracket fixings, or masonry foundation slab for builder. Set up calorifier on brackets or slab, level up, fit all mountings, and clean off joints, all ready for steam and condensate piping and water connections.

A. Ground floor plant room.
B. Basement plant room.

Hours, fitter and mate

H.W.S. storage calorifiers				Primary, or non-storage calorifiers for H.W.H. or H.W.S.			
Capacity		Hours		Duty		Hours	
Litres	Gallons	A	B	*kW	Btu/h	A	B
450	100	11·5	12·5	59	200 000	6·5	7·5
680	150	13·0	14·0	117	400 000	8·5	9·5
910	200	14·5	15·5	147	500 000	10·5	11·5
1 365	300	18·5	20·0	234	800 000	15·0	16·5
1 820	400	26·0	28·0	293	1 000 000	17·0	19·0
2 270	500	32·0	34·0	586	2 000 000	19·5	21·5
				879	3 000 000	22·0	24·0

* To nearest whole number.
Note: When applicable, apply variations as for Section 20.

Section 22

WATER TO WATER HEATING CALORIFIERS

Off load at site, one calorifier, altitude gauge, safety valve, thermometer, empty cock, and steel brackets or supports. Move to fixing position in plant room of new building. Mark out bracket fixings, or masonry foundation slab for builder. Set up calorifier on brackets or slab, level up, fix mountings, and clean off joints, all ready for piping connections.

A. Ground floor plant room.
B. Basement plant room.

SECTION 22 (*Contd.*)

Calorifier heating capacity kW	Calorifier heating capacity Btu/h	Hours, fitter and mate	
		A	B
59	200 000	8·0	9·0
117	400 000	10·0	11·0
147	500 000	12·0	13·0
234	800 000	16·0	17·0
293	1 000 000	20·0	21·5
586	2 000 000	22·0	24·0
879	3 000 000	25·0	27·0

Note: Where applicable, apply variations as for Section 20.

Section 23

ELECTRICALLY-DRIVEN CENTRIFUGAL PUMPS

Off load at site, one pump, counter flanges or unions, holding-down bolts, 2 stop-valves, drip pipe and fittings, brackets, and gland packing. Move to fixing position in plant room of new building. Mark out bracket fixings or masonry base for builder. Set up pump on base or brackets, bolt down and level, cut and screw inlet and discharge pipes, fit valves and unions or flanges, and connect pump to system. Fit drip pipe, pack pump gland, fill lubricators, clean joints, test for rotation after wiring by electrical contractor, and leave pump in full working order.

Note: Flanged connections 65 mm (2½ in) and above, unions 50 mm (2 in) and below.
A. Ground floor plant room.
B. Basement plant room.

Pump size mm	in	Motor and pump horizontal direct coupled		Pump and motor separate open belt drive		Fullway pump fitted to pipe line		Sump pump on wall
		A	B	A	B	A	B	
20	¾	4·5				3·5	4·0	2·0
25	1	4·5	5·0	5·5	6·0	3·5	4·0	2·5
32	1¼	6·5	7·0	7·5	8·0	4·5	5·0	3·0
40	1½	8·5	9·0	9·5	10·0	6·0	6·5	3·5
50	2	10·5	11·0	11·5	12·0	7·5	8·0	4·0
65	2½	12·5	13·0	13·5	14·0	8·5	9·0	
80	3	14·5	15·5	16·0	16·5	10·5	11·5	
100	4	16·5	17·5	18·0	18·5	12·5	13·5	

Section 24

MISCELLANEOUS ITEMS OF FIXING

Off load, move to fixing position, and fix the following:

Job	Hours, fitter and mate
Tool rack on boiler room wall	1·5
Starter, switch, time control, control box, draught gauge	1·5
Room thermostat	1·0
Outdoor thermostat	3·0
Duct thermostat	2·0
Capillary tubing per metre run	0·25
Small bore tubing for gauges, etc. per metre run	0·15
Draught regulator or stabiliser, fix and set	2·0
Radiator wall shield	2·0
Towel rail to wall	4·0
Towel rail to floor	2·0
H.W.S. mixing valve to wall	1·5
H.W.S. shower rose or spray fitting to wall	1·5
Hit and miss ventilator in wall	1·5
Louvre ventilator, and quadrant	3·0

Section 25

OPEN TOP GALVANISED STEEL WATER STORAGE OR FEED AND EXPANSION TANKS

Off load at site, one tank and lid, and ball valve, and move to fixing position in tank room or roof space of new building. Set tank on supports prepared by builder, pack up and level, and fit ball valve and sheet steel lid.

A. Fixed at up to 6 m from ground floor.
B. Fixed at up to 15 m from ground floor.
C. Fixed at up to 30 m from ground floor.

Tank capacity litres	Tank capacity gallons	Hours, fitter and mate		
		A	B	C
Up to 91	Up to 20	3·0	3·5	4·0
230	50	4·0	5·0	5·5
455	100	5·5	7·0	8·0
910	200	9·0	11·0	12·5
1 365	300	10·0	12·5	14·0

SECTION 25 (*Contd.*)

Tank capacity litres	Tank capacity gallons	Hours, fitter and mate		
		A	B	C
1 820	400	14·0	16·5	18·5
2 270	500	18·0	23·0	25·5
2 730	600	23·0	29·0	32·0
3 400	750	28·0	35·0	39·0
4 500	1 000	36·0	45·0	50·0
6 800	1 500	42·0	53·0	59·0
9 100	2 000	54·0	68·0	75·0

Variations
For work in restricted roof space add 30 per cent.
For work on domestic system in occupied house add 20 per cent.

Section 26

RECTANGULAR STEEL, FUEL OIL STORAGE TANKS

Off load at site, one tank, move to tank room on ground floor of new building. Set tank on supports prepared by builder, level and line up ready for filling, venting, supply to burners, and for sludging.

Tank storage capacity litres	Tank storage capacity gallons	Hours, fitter and mate
455	100	2·0
1 250	275	3·5
2 270	500	11·5
2 800	625	15·0
3 400	750	21·0
4 500	1 000	24·0
5 700	1 250	26·0
6 800	1 500	28·0
7 900	1 750	31·0
9 100	2 000	36·0

Variations
For cylindrical tanks add 10 per cent.
For work in occupied building with unrestricted space and good access add 10 per cent.

Section 27

SHEET STEEL DUCTS
(Weights as Standard Specification, Clause 21, Section B)

Off load at site, and move ducts and brackets to fixing positions on ground floor of new building. Arrange scaffolding or other fixing platform. Mark out duct runs and bracket fixings. Fix hangers and brackets. Trim ductwork, joint and erect straight runs at a maximum height of 6 m from floor. Mark out for, trim and fix branches and bends. Remove scaffolding and clean up.

A. Straight runs per metre.   **B.** Fit bend, branch, damper, or connecting piece.

Duct size, Metric		Duct size, Imperial		Hours, fitter and mate	
Rectangular mm	Circular equivalent mm	Rectangular in	Circular equivalent in	A	B
150 × 100	134	6 × 4	5·4	0·30	0·50
250 × 100	171	10 × 4	6·8	0·40	0·60
200 × 150	190	8 × 6	7·6	0·40	0·60
250 × 150	212	10 × 6	8·4	0·50	0·80
200 × 200	220	8 × 8	8·8	0·50	0·80
400 × 150	264	16 × 6	10·5	0·65	1·15
300 × 200	269	12 × 8	10·7	0·60	1·10
250 × 250	275	10 × 10	11·0	0·60	1·10
300 × 250	301	12 × 10	12·0	0·65	1·15
500 × 200	341	20 × 8	13·6	0·85	1·35
600 × 200	371	24 × 8	14·8	0·95	1·45
500 × 250	385	20 × 10	15·4	1·05	1·50
600 × 250	419	24 × 10	16·8	1·10	1·60
500 × 300	424	20 × 12	17·0	1·10	1·60
400 × 400	440	16 × 16	17·6	1·15	1·70
700 × 300	496	27 × 12	19·5	1·30	2·15
600 × 400	537	24 × 16	21·5	1·50	2·20
700 × 400	578	27 × 16	22·7	1·60	2·30
600 × 500	603	24 × 20	24·0	2·00	2·50
700 × 500	650	27 × 20	25·5	2·10	2·60
700 × 600	713	27 × 24	28·0	2·20	2·70
700 × 700	761	27 × 27	29·7	2·30	3·25
800 × 600	770	32 × 24	30·5	2·30	3·25
800 × 700	824	32 × 27	32·3	2·75	3·50
800 × 800	881	32 × 32	35·2	3·00	4·00

Variations
For each 3 m above 6 m of fixing height add 15 per cent.
For each floor above or below ground floor add 10 per cent.
For occupied buildings with good access add 40 per cent.

The duct sizes in Section 27 table, are in accordance with the H.V.C.A. metric publication DW/112, standard range of rectangular ducts and fittings. (See Appendix A, recommended publications.) This publication gives a rationally reduced range of duct sizes to enable advantage to be taken of more economic production methods, to simplify design, accelerate site work, and also enable quicker delivery to be made when additional ductwork may be called for due to site variations.

Section 28

ELECTRICALLY-DRIVEN CENTRIFUGAL FANS AT LOW LEVEL IN PLANT ROOM

Off load at site, fan and motor and move to fixing position in plant room on ground floor of new building. Set out foundation base, and mark out holding-down bolt holes for builder. Set machine on base, level and line up with ductwork, grout in holding-down bolts, bolt down, fill lubricators, check for rotation after wiring by Electrical Contractor, and leave machine in full running order. Clear away all lifting tackle and packing material used for erection.

A. Fan with direct coupled motor.
B. Fan with separate motor and open drive.
C. Add for each floor above ground floor.

Fan size mm	Fan size In	Hours, fitter and mate		
		A	**B**	**C**
381	15	10	12	0·5
508	20	12	14	1·0
762	30	18	22	1·5
1016	40	22	28	2·0
1270	50	27	32	2·5
1524	60	36	42	3·0

Variations
For fixing on platforms up to 4·6 m above floor level add 10 per cent.
For work in occupied building with good access add 20 per cent.

Section 29

AIR HEATERS
Hot Water or Steam Gilled Tube Pattern

Off load at site, one air heater and move to fixing position at low level in ground floor plant room of new building. Mark out support or base for builder. Move heater on to base, level and line up with mating flanges, make joints and bolt up. Remove lifting tackle and clean joints.

Note: Steam or hot water connections are not included.

A. Heater with **single** row of tubes.
B. Heater with **two** rows of tubes.
C. Heater with **three** rows of tubes.

SECTION 29 (*Contd.*)

Heater size mm	Heater size in	Hours, fitter and mate		
		A	B	C
300 × 300	12 × 12	4·0	4·5	5·0
300 × 450	12 × 18	4·5	5·0	5·5
450 × 450	18 × 18	5·0	5·5	6·0
450 × 600	18 × 24	5·5	6·0	7·0
600 × 600	24 × 24	6·5	7·0	8·0
600 × 900	24 × 36	8·0	8·5	9·5
750 × 900	30 × 36	9·0	9·5	10·0
900 × 900	36 × 36	10·0	10·5	11·0
750 × 1 200	30 × 48	10·5	11·0	12·0
900 × 1 200	36 × 48	11·5	12·0	13·0
1 200 × 1 200	48 × 48	13·5	14·5	16·0
1 200 × 1 500	48 × 60	15·5	16·5	18·0
1 500 × 1 500	60 × 60	18·0	19·0	20·5
1 500 × 1 800	60 × 72	20·0	21·5	23·5
1 800 × 1 800	72 × 72	22·0	24·0	26·5

Variations
For work in occupied buildings with good access add 20 per cent.
For work on each floor above or below ground floor add 10 per cent.
For fixing on platforms up to 4·6 m above floor level add 15 per cent.

Section 30

AIR FILTERS
Cotton Wool or Manual Viscous Types

Off load at site, one air filter and move to fixing position at low level in ground floor plant room of new building. Mark out support or base for builder. Move heater on to base, level and line up with mating flanges. Make joints and bolt up. Remove lifting tackle, and clean joints. Fit filtering elements ready for use.

A. Filter connected into ductwork.
B. Filter outlet bolted to ductwork and inlet built into outside wall by builder.

Filter size (approx.) mm	Filter size (approx.) in	Hours, fitter and mate	
		A	B
600 × 900	24 × 36	6·0	4·5
800 × 900	32 × 36	6·5	5·0
1 000 × 900	40 × 36	8·0	5·5
1 000 × 1 350	40 × 54	9·0	6·0
1 200 × 1 350	48 × 54	10·0	6·5
1 400 × 1 350	56 × 54	11·0	7·0
1 600 × 1 350	64 × 54	12·0	8·0
1 800 × 1 350	72 × 54	13·0	9·0

Note: Variations for Section 29 apply.

Section 31
METAL INLET OR OUTLET GRILLES

Off load at site, one grille, register or diffuser and move to fixing position on ground floor of new building. Set up scaffold or ladder when necessary, mark out for builder for fixing wood frame, or Rawlplug wall for fixing. Fit grille into position, and screw or bolt up. Remove scaffold or ladder.

A. Fixed to wood frame provided by builder.
B. Fixed to wall with Rawlplugs or similar fixing.

| Grille size mm | Grille size in | *Hours, fitter and mate* | | | |
| | | **A** | | **B** | |
		At low level	At high level	At low level	At high level
300 × 300	Up to 12 × 12	1·00	1·5	1·50	2·25
600 × 600	24 × 24	1·50	2·0	2·00	3·00
750 × 900	30 × 36	1·75	2·5	2·25	3·25
900 × 1 200	36 × 48	2·25	3·0	3·00	4·00
1 200 × 1 200	48 × 48	2·50	4·0	3·50	5·00
1 200 × 1 500	48 × 60	3·00	5·0	4·00	6·00

Variations
For each floor above ground floor add 5 per cent.
For work in occupied building add 30 per cent.

Section 32
PROPELLER FANS

Off load at site, one propeller fan and move to near fixing position on ground floor of new building. Set up scaffold. Mark off for builder's work. Haul up fan to high level fixing point, and bolt up to fixing board or frame ready for wiring. Check fan rotation after wiring by electrician.

A. Fixed at high level on ground floor.
B. Add for each floor, above or below ground floor.

| Fan size mm | Fan size in | *Hours, fitter and mate* | |
		A	**B**
225	9	5·0	0·5
375	15	5·5	0·5
450	18	6·0	0·5
525	21	6·5	0·5
600	24	7·0	0·75
750	30	7·5	0·75
900	36	8·0	0·75
1 050	42	9·0	1·00
1 200	48	10·0	1·00

Variation
For work in occupied building add 30 per cent.

Section 33

OUTLET COWLS, LOUVRE SHUTTERS, BUTTERFLY SHUTTERS OR OUTLET BENDS
(For Outside Wall Fixing in conjunction with Propeller Fan)

Off load at site, one cowl or one shutter, and move to near fixing position on ground floor. Set up ladder or scaffold. Mark off, plug wall or fix rag bolts. Haul up cowl or shutter, fit to bolts, tighten up, and check that shutter is operating correctly. Remove ladder or scaffold.

A. With four fixing lugs.   **B.** With fixing flange, eight holes.

Fan size mm	Fan size in	Hours, fitter and mate	
		A	B
225	9	4·5	5·5
360	12	4·5	5·5
375	15	4·5	5·5
450	18	5·0	6·0
525	21	5·0	6·0
600	24	5·0	6·0
750	30	5·5	6·5
900	36	6·0	7·0
1 050	42	7·0	8·0
1 200	48	7·5	8·5

Variation
For each 3 m height above 7·5 m add 20 per cent.

Section 34

TESTING AND SUPERVISION

Hours to be included for labour (1 fitter and 1 mate), for supervision at site; filling up; venting; balancing; testing heating and ventilating systems; clearing up after completion of erection and operating instructions to Owner.

A. Estimated hours for erection of complete installation.
B. Hours to be added (Constant plus percentage of estimated hours).

A	B
Up to 100	$30\% \times A$
101 to 150	$21 + (10\% \times A)$
151 to 300	$24 + (\ 8\% \times A)$
301 to 500	$30 + (\ 6\% \times A)$
501 to 2 000	$35 + (\ 5\% \times A)$
2 001 to 3 000	$55 + (\ 4\% \times A)$
3 001 to 4 000	$115 + (\ 2\% \times A)$

Example. Estimated hours for erection = 360 hours.
Time to add for supervision, testing, etc.:

$$30 \text{ hours} + \frac{6 \times 360}{100}$$

$$= 30 \text{ hours} + 21 \cdot 6 \text{ hours}$$
$$= 51 \cdot 6 \text{ hours.}$$

Section 35

DISMANTLING WORK

Dismantle and remove from site existing equipment and installations, including loading on wagon ready to leave site.

Item	Multiplier
Steel pipework	{ Section 1 × 0·5 { Section 4 × 0·5
Light gauge copper pipework	Section 2 × 0·3
Cast iron pipework	Section 6 × 0·3
Radiators	Section 8 × 0·6
Convectors	Section 10 × 0·5
Radiant panels	Section 13 × 0·7
Radiant panels	Section 14 × 0·7
Unit heaters	Section 15 × 0·7
Unit heaters	Section 16 × 0·7
Sectional heating and H.W.S. boilers	Section 17 × 0·6
Boiler mountings	Section 18 × 0·25
Smoke pipe and fittings	Section 19 × 0·25
Hot water storage cylinders	Section 20 × 0·6
Steam calorifiers (storage)	Section 21 × 0·6
Non-storage calorifiers	Section 21 × 0·6
Heating (water) calorifiers	Section 22 × 0·6
Pumps	Section 23 × 0·3
Miscellaneous items	Section 24 × 0·3
Water tanks	Section 25 × 0·5
Fuel oil tanks	Section 26 × 0·6
Steel ductwork	Section 27 × 0·5
Centrifugal fans	Section 28 × 0·5
Air heaters	Section 29 × 0·5
Air filters	Section 30 × 0·5
Air grilles	Section 31 × 0·3
Propeller fans	Section 32 × 0·4
Outlet cowls, etc.	Section 33 × 0·4

Note: When using the multiplying factors for estimating labour for dismantling and taking out old installations, the variations for height, etc., given for each section must be taken into account in the same way as for the erection of new work.

Section 36

FORCED WARM AIR DOMESTIC CENTRAL HEATING
Light Gauge Ductwork

Off load at site, and move ladders, scaffolding, ducts, fittings, brackets, insulating materials and tools, to fixing position on the ground floor of a new unoccupied building. Mark out duct runs and fixings. Fix hangers and brackets. Trim ductwork, joint and erect straight runs. Mark out for, trim and fix branches, bends, etc.

Remove ladders or scaffolding, and clean up.

A. Diameter of duct or length of longest side (approx.).
B. Straight runs per metre.
C. Fit bend, branch, damper, starting elbow, starting collar, transition piece, flexible connection, floor or wall register.
D. As for **B**, insulated with 25 mm (1 in) thick rigid or flexible duct insulation placed on adhesive, with joints sealed with adhesive tape.
E. As for **C** insulated as **D**.

A		*Hours, fitter and mate*			
mm	*in*	**B**	**C**	**D**	**E**
125	5	0·30	0·50	0·45	0·75
150	6	0·33	0·55	0·50	0·80
200	8	0·36	0·60	0·55	0·90
250	10	0·48	0·70	0·63	1·00
300	12	0·55	0·85	0·80	1·25
350	14	0·65	1·00	0·90	1·50
400	16	0·80	1·30	1·20	1·90
450	18	0·90	1·40	1·35	2·10
500	20	1·10	1·60	1·70	2·40
600	24	1·50	1·75	2·25	2·65
750	30	1·80	2·50	2·50	3·75

Variations

For work in roof space add 20 per cent.
For work on 1st floor add 5 per cent.
For work concealed in floor void add 15 per cent.
For work in occupied house add 30 per cent.

Gas or Oil Fired Air Heaters

Off load heater at site, with all mountings, pipes, valves, outlet and return warm air duct connecting pieces, flue pipe, and fittings, electrical conduit, fittings, and control equipment and wiring.

Place heater on prepared concrete plinth in new unoccupied building and make necessary gas (or oil) connection to fuel supply and to burner. Electrical connections in conduit to burner, heater fan and controls, and to the electricity supply.

Connect the outlet and return air ductwork systems to the heater and test.

Heater output kW	Heater output Btu/h	Hours, fitter and mate	Hours, electrician and mate	Total hours fixing
8·79	30 000	9·0	5	14·0
11·72	40 000	10·5	5	15·5
14·66	50 000	12·5	5	17·5

Variation
For work in occupied house add 20 per cent.
Note: Gas or oil supply not more than 6 m from heater.
Electricity supply not more than 9 m from heater.

Electric Air Heaters for Fitting to Ductwork System

Off load at site, electric air heater, nuts, bolts, washers, conduit, wiring, and control gear. Move to fixing position on ground floor of new unoccupied building.

Mark out for supports or hangers. Move on to supports or hangers, level and line-up with mating flanges, make joints and bolt up. Remove lifting tackle and clean joints.

Wire in heavy gauge conduit from point of supply (maximum 9 m) through isolating switch-fuse unit, control thermostat and contactor to heater terminal box, test, and clean up.

Heater loading kW at 240 volts	Hours, fitter and mate	Hours, electrician and mate	Total hours fixing
12 to 18	4·5	14·5	19·0
24 to 36	5·5	16·0	21·5

Variation
For work in occupied house add 15 per cent.

Off-Peak Electric Warm Air Central Heating Units

Off load at site, air heater assembly, including cabinet, fan unit, plenum chamber, outlet warm air and return air inlet duct connecting pieces, all control and wiring equipment, and move to fixing position in a new unoccupied building.

Assemble heating unit, make all electrical connections to heater, fan, thermostat and time controls and to the electricity supply (maximum 9 m).

Connect the warm air outlet and return air ductwork systems to the heater, clean up and test.

A. Heater fixed in room on ground floor with unrestricted working space.
B. Heater fixed in space under stairs.
C. Heater fixed in basement with unrestricted working space.

Heater loading kW at 240 volts	Hours, electrician and mate		
	A	B	C
9	20	23	22
12	23	26	25

Variation
For work in occupied house add 15 per cent.

Section 37

ELECTRIC DOMESTIC SPACE HEATING

Wiring in Screwed Conduit

Off load at site, conduit and couplers, bench and tools, and move to fixing position on ground floor of new unoccupied building.

Arrange and mark out conduit runs from supply position to heater switch points.

Fix conduits only, connect to point of supply, draw in cables, clean up and test.

A. Surface work on timber.
B. Surface work on brick or concrete.
C. Sunk work including chasing in brickwork.
D. Sunk work including chasing in concrete.

Conduit size mm	Conduit size in	Electrician and mate, Hours per metre run			
		A	B	C	D
16	$\frac{5}{8}$	0·30	0·33	0·48	0·51
20	$\frac{3}{4}$	0·36	0·39	0·54	0·57
25	1	0·42	0·45	0·66	0·75

Variations
For work on 1st floor add 7 per cent.
For work in roof space add 15 per cent.
For work in occupied house add 30 per cent.

Conduit Fittings

Off load at site, move to working position on ground floor of new unoccupied building.

Fix conduit fittings up to and including size 25 mm (1 in).

Description	Hours, electrician and mate
Elbow	0·20
Tee	0·30
Through tee	0·20
Four Way tee	0·35
Boxes	
Through or angle	0·20
Four Way	0·35
Tee box	0·30
Back outlet or terminal	0·20
Mounting Box	0·25

Variation
As for conduit.

Wiring in Butyl and C.P. Rubber or P.V.C.

Off load at site, cables, buckle clips, and screws, tools and bench. Move to working position on ground floor of new unoccupied building. Install wiring, connect to point of supply, clean up and test.

Cable size No. and dia. (mm) of wires	Electrician and mate, hours per metre run					
	A	B	C	D	E	F
1/1·38	0·21	0·22	0·18	0·19	0·22	0·23
1/1·78	0·21	0·22	0·18	0·19	0·22	0·23
7/0·85	0·22	0·28	0·19	0·24	0·23	0·24
7/1·04	0·22	0·28	0·19	0·24	0·23	0·24
7/1·35	0·24	0·29	0·21	0·25	0·24	0·25
7/1·70	0·30	0·31	0·27	0·26	0·25	

A. Two single core cables. **D.** Flat three core cable.
B. Three single core cables. **E.** Two cables into wood or metal channelling.
C. Flat twin cable. **F.** Three cables into wood or metal channelling.

Variations
For work on 1st floor add 7 per cent.
For work in roof space add 15 per cent.
For work in occupied house add 30 per cent.

Electrical Fittings and Accessories

Off load at site, switches, fuseboards, thermostats, socket outlets, etc., tools and bench. Fix and wire to installed cable system, clean up and test.

Description	Hours, electrician and mate
Domestic Consumers' Unit; or Switch-Fuseboard	
15 amp 6 way	4·5
15 amp 8 way	5·5
15 amp 10 way	6·5
15 amp DP contractor	2·0
15 amp switch on wall	1·0
15 amp ceiling switch	1·5
13 amp fused spur unit	1·0
13 amp or 15 amp socket outlet	1·0
20 amp room thermostat	1·0
15–30 amp switch fuse T.P.N.	2·5
15–30 amp SP and N or DP switch fuse	1·5
45–60 amp SP and N or DP switch fuse	2·0
Time switch	2·5
Immersion Heater (Domestic Hot Water Cylinder)	3·0

Variations
For work on 1st floor add 7 per cent.
For work in roof space add 15 per cent.
For work in occupied house add 30 per cent.

Electric Thermal Storage Radiators

Off load at site, electric thermal storage radiators, cables, conduit, conduit fittings, bench and tools. Move to fixing positions in occupied building, install wiring from distribution board of consumers' unit at the meter position to heater switches, clean up and test.
A. Wiring in screwed concealed conduit.
B. Wiring in concealed P.V.C.

Heater loading kW at 240 volts	Hours, electrician and mate							
	Fully assembled heater				Heater assembled on site			
	Fixed on ground floor		Fixed on first floor		Fixed on ground floor		Fixed on first floor	
	A	B	A	B	A	B	A	B
1·25–1·50	6·0	5·0	8·5	6·5	7·5	6·5	9·0	8·0
2·00–2·50	6·5	5·5	9·0	7·0	8·5	7·5	11·0	9·0
2·75–3·00	7·0	6·0	9·5	7·5	9·5	8·5	12·0	10·0

Note: This table does not include time for fixing and connecting to room thermostats, and time switch.

For these time rates refer to table at top of page 158.

Solidly Embedded Electric Floor and Ceiling Heating Cable Installations in Houses and Bungalows

Off load at site, cables, wiring accessories, jigs and spacers, and move to fixing position in building under construction.

Erect fixing platform for ceiling heating work, mark out heating panel, and staple heating cables to fire proofed plaster board fixed by builder.

For floor warming, mark out heating panel, fix cable jigs or spacers, lay heating cables on surface of sub-floor.

Set up monitor, and supervise during screeding of floor, or plastering of ceiling to check possible damage to heating cables.

Remove cable laying jigs as necessary during screeding.

Connect heating cable tails at terminal box in room to be heated.

Type of heating	Hours, electrician and mate per m² of heating panel surface
Solidly embedded floor warming cables	1·3
Solidly embedded ceiling heating cables	1·1
Prefabricated ceiling heating elements (e.g. I.C.I. Flexel) including fixing of fibre-glass insulation backing	2·2
Note: For circuit wiring see pp. 156 and 157	

Variations

For 1st floor work add 7 per cent.

For ceiling heating in occupied houses add 20 per cent.

Section 38

SMALL BORE AND MICRO BORE DOMESTIC CENTRAL HEATING

The information given in this section applies particularly to small bore installations in houses and bungalows and is supplementary to Sections 2, 3, 5, 8, 9, 10, 11, 12, 17, 18, 19, 23, 24, 25, 26, and 35.

Packaged, Prewired Gas and Oil Fired Boiler Units for Domestic Small Bore Hot Water Space Heating, and Hot Water Supply

Off load at site, packaged boiler unit and flue connections, and move to fixing position in ground floor room of new unoccupied building. Mark out boiler base for builder, place unit in position. Connect to prepared flue, or (in case of gas) to balanced draft terminal. Connect to gas or oil supply from oil tank or gas meter assumed to be 6 m from boiler unit.

Electrician to fix isolating switch-fuse, and to wire from house meter position (assumed to be 9 m from boiler unit) and connect at supply position and boiler unit terminal box.

Clean up, and test gas or oil supply and controls, and test electrical installation.

Boiler output rating kW	Boiler output rating Btu/h	Hours, fitter and mate	Hours, electrician and mate	Hours fixing. To be added where pump is not included in the packaged unit	
				Fitter and mate	Electrician and mate
11·72	40 000	5·25	4·5	1·25	2·0
14·66	50 000	5·25	4·5	1·25	2·0
17·57	60 000	6·25	4·5	1·25	2·0
20·52	70 000	7·00	4·5	1·25	2·0
*14·66	5 000	3·00	2·0		

*Finned tube wall mounted gas boiler.

Variations
For work in occupied house add 10 per cent to mechanical labour, and 15 per cent to electrical labour.

Small Solid Fuel Boilers

Off load at site for installation in house or bungalow under construction and move to fixing position. Mark out boiler base for builder. Place unit in position and connect to flue. Fit safety valve, altitude gauge, thermometer, and empty cock. In the case of automatic boilers, wire to fan and thermostat from house meter position 9 m from boiler unit, clean up and test.

Boiler output rating kW	Btu/h	Hours, fitter and mate	Add for automatic boiler Hours, electrician and mate
10·23	35 000	4·00	4·5
11·72	40 000	4·00	4·5
13·19	45 000	4·50	4·5
14·66	50 000	4·50	4·5
17·58	60 000	5·00	4·5
20·52	70 000	5·55	4·5

Variations
For work in occupied houses add 10 per cent to mechanical labour, and 15 per cent to electrical labour.

Open Fire Back Boiler Unit for Domestic Space Heating and Hot Water Supply

Off load at site, back boiler unit, and move to fixing position in ground floor room of new unoccupied building.

Set on level base in builder's fireplace opening, and connect space heating flow and return pipes and primary hot water flow and return pipes, and continue these through chimney breast brickwork ready for connecting to the respective circuits.

Build in boiler with damper box and damper blade, form outlet flue with lining bricks. Fill in solid with weak mixture of rubble and cement or lime mortar. Fix the fire surround and front hearth, clean up, and leave to dry out.

Boiler size	Boiler size	Hours, fitter and mate	Hours, bricklayer and mate
400 mm	16 in.	7·0	9·0
450 mm	18 in	7·5	10·5

Variations
When fixed in occupied house, and replacing existing traditional type back boiler, add 25 per cent to engineering labour and 30 per cent to building labour.

Domestic Small Bore Circulating Pumps (Direct Coupled Electrically Driven)

Off load at site, connect to system, wire up, and test.

Pump size mm	in	Hours, fitter and mate	Hours, electrician and mate
19	$\frac{3}{4}$	2·0	3·5
25	1	2·5	3·5

Variations
For work in occupied house add 10 per cent to mechanical work, and 15 per cent to electrical work.

Domestic Hot Water Cylinder (Indirect Systems)

Off load at site one cylinder, one pair of brackets or other supports. Move to fixing position on ground floor of new building, mark out bracket fixings for builder, set cylinder on brackets or supports, and prepare tappings for fixing pipe connections.

Cylinder capacity		House under construction Hours, fitter and mate		Occupied house Hours, fitter and mate
			Direct cylinder, including cutting hole for 65 mm boss and fitting micro-bore heating coil	Prepare for and fit microbore heating coil to existing cylinder
litres	gallons	Indirect cylinder		
100	22	3·00	4·50	3·0
114	25	3·00	4·50	3·0
123	27	3·50	5·00	3·0
145	32	4·00	5·00	3·0
170	37	4·50	5·50	3·0
214	47	5·00	6·00	3·0

Variations
For work in occupied house add 10 per cent to the times in Columns 3 and 4. Add 5 per cent each floor above ground floor.

Light Gauge Copper Pipes (B.S. 2871) and Capillary Fittings fixed on the wall surface on the ground floor in new buildings

Off load at site, pipes, fittings, valves, floor plates, and brackets, and move to fixing position in house or bungalow.

Arrange pipe runs, mark out for and fix brackets, measure for and cut pipe to required length, solder joints, clean off, and secure pipework to brackets.

Nominal pipe size (bore)		Hours, fitter and mate			
				Per pipe bracket	
mm (o.d.)	in (i.d.)	Per metre of pipe	Per fitting valve etc.	Screw-on	Build-in
12	$\frac{3}{8}$	0·10	0·25	0·08	0·25
15	$\frac{1}{2}$	0·11	0·30	0·08	0·25
22	$\frac{3}{4}$	0·13	0·30	0·08	0·25
28	1	0·15	0·35	0·08	0·25
35	$1\frac{1}{4}$	0·18	0·40	0·10	0·30

Variations
For work under floor boards add 30 per cent.
For each floor above ground floor add 5 per cent.
For work in occupied houses add 20 per cent.

Microbore Copper Pipes and Fittings

Off load pipe and fixing clips at site, and move to fixing position on the ground floor of new building under construction.

Arrange pipe runs from plan, level and mark out clips for surface fixing, and fix brackets or clips and sound insulation pads. Cut pipe to required length, form bends and or off-sets, fit couplings as required. Fix pipework to skirtings, or walls, and clean up after fixing.

Off load fittings and valves at site, and move to fixing position. Measure up for fixing, cut pipe and fix, and clean up joints.

Hours, fitter and mate

Pipe size mm o.d.	Pipe per m run	Pipe fittings each	Form bend	Form off-set	Connect manifold to system			
					Main circuit 22 mm (¾ in) B.S. 2871		Main circuit 28 mm (1 in) B.S. 2871	
					No. of outlets	Hours	No. of outlets	Hours
6 and 8	0·10	0·30	0·25	0·60	4	1·50	6	2·10
10	0·11	0·33	0·30	0·70	6	2·00	8	2·60
12	0·13	0·35	0·35	0·80	8	2·50	10	3·10

Variations
For work under floor boards add 30 per cent.
For each floor above ground floor add 5 per cent.
For work in occupied houses add 20 per cent.

Steel Radiators for Small Bore and Microbore Domestic Heating Systems

Small Bore System
Off load at site, radiators, wall brackets, radiator valves (single or twin entry), fittings, and tube for connections, and move to ground floor of new building. Mark out for, and fix radiator wall brackets, fit radiator to brackets, fit radiator valves, and fit tees in mains. Measure up for, cut pipe, and fit pipe connections to mains and radiator.

Microbore System

Off load at site, radiators, wall brackets, radiator valves (single or twin entry). Mark out for and fix wall brackets, fit radiator to brackets, fit two single entry or one twin entry radiator valve to radiator, and connect flow and return pipes to radiator.

Small bore		Hours, fitter and mate				Microbore
Size of mains		Size of small bore radiator connections				Hours, fitter and mate
mm (o.d.)	in (i.d.)	8 mm ($\frac{1}{4}$ in)	12 mm ($\frac{3}{8}$ in)	15 mm ($\frac{1}{2}$ in)	22 mm ($\frac{3}{4}$ in)	Connections 6 mm, 8 mm, or 10 mm
12	$\frac{3}{8}$	2·25	2·25			2·00
15	$\frac{1}{2}$	2·50	2·50	2·75		—
22	$\frac{3}{4}$		2·75	3·00	3·25	—
28	1			3·25	3·50	—

Variations

For work in occupied houses add 20 per cent.
For radiators requiring 3 or 4 brackets add 0·35 h or 0·7 h.
For each floor above ground floor add 5 per cent.

Section 39

THERMAL INSULATION TO PIPEWORK (STEEL)

Off load at site, magnesia, glass fibre, or other type of sectional insulation for pipes and fittings, and move to fixing position on ground floor of new building.

Trim as required, and apply the insulation to pipework.

Clean off after fixing and leave all parts tidy.

A. Pipes per metre.
B. Tee, elbow, bend, or other fitting.

Pipe bore (Steel: B.S.1387)		Hours, fitter and mate	
mm	in	A	B
15, 20 and 25	$\frac{1}{2}$, $\frac{3}{4}$ and 1	0·15	0·15
32 and 40	$1\frac{1}{4}$ and $1\frac{1}{2}$	0·18	0·18
50 and 65	2 and $2\frac{1}{2}$	0·21	0·21
80 and 100	3 and 4	0·27	0·27

Note: This table may also be used for B.S.2871 copper tubes.

Variations
As for Section 1.

Thermal Insulation to Pipework. Microbore, Small Bore, and Domestic Hot Water Systems

Off load at site, mineral wool pipe wrapping strip and move to new house or bungalow ready for fixing. Unwrap and apply to (A) pipes in floor spaces, (B) pipes in roof space, and (C) pipes in rooms, e.g. vertical risers which are to be cased-in.

Pipe bore Copper B.S.2871 and steel B.S.4127		Hours, fitter and mate		
mm (o.d.)	*in* (i.d.)	**A**	**B**	**C**
15 and 22	$\frac{1}{2}$ and $\frac{3}{4}$	0·17	0·15	0·12
28	1	0·18	0·16	0·13
35	$1\frac{1}{4}$	0·21	0·18	0·15
Microbore tube outside diameter mm				
6 and 8		0·13	0·11	0·10
10 and 12		0·15	0·13	0·11

Variation
For work in occupied house, add 20 per cent.

5

PREPARING THE TENDER

THE BILL OF QUANTITIES

The foundation of the tender is the Priced Bill of Quantities, in which each item of equipment and unit of materials is listed and priced, together with the estimated cost of labour for erection.

Net prices at which the Contractor may purchase equipment and materials are obtained from the manufacturers' trade lists, or, as explained in Chapter 3, by special quotation for certain machines and those pieces of equipment which are not usually listed.

The final estimate of labour cost is made up of the primary charge for the actual work of erection, testing, and also supervision by the chargehand, or walking foreman, plus secondary costs which include the employer's National Insurance and Selective Employment Tax, Redundancy Fund, the Holidays with Pay Credits, and Sickness and Accident premiums respectively.

Where it is known at the time of tendering that certain work can only be carried out during periods outside the normal working day, or at weekend, overtime payment must be included in the labour estimate.

Other secondary labour costs vary according to the nature of the job and its distance from the firm's headquarters or from the branch handling the job. These are lodging allowances; fares; travelling time; dirty money for exceptionally dirty work or work under abnormal conditions; and danger money when working at heights of 6 m or more from the nearest fixed flooring or fixed scaffolding to the actual work.

The amount of these extra payments (wages and working week) and the circumstances governing their payment are laid down in the National Agreement for the Heating, Ventilating and Domestic Engineering industry, and the estimator will apply the rates current at the time of tendering.

A TYPICAL EXAMPLE

The example which follows shows the preparation of a tender for providing and fixing the low pressure gravity hot water heating

BILL OF QUANTITIES FOR HEATING HALL

Sheet No. 1   *Date:* 25.6.71

Main Hall		Materials							Labour				
		Unit net price						Hours fixing					
		£	p	£	p	£	p		£	p	£	p	
1	70 litre galvanised feed and expansion tank and cover	3	75	3	75			4·00	4	0			
1	15 mm ball valve	1	45	1	45								
4·6 m	32 mm medium galvanised steel tube for overflow		33	1	52			0·97		97			
2	pipe hangers		75	1	50								
1	32 mm galvanised M.I. bend		41		41	8	63	0·55		55	5	52	
6 m	25 mm medium galvanised steel tube for F and E pipe		26	1	56			1·14	1	14			
2	25 mm build-in pipe brackets		10		20								
1	25 mm G.M. gate valve	1	30	1	30			0·50		50			
2	25 mm M.I. galvanised bends		28		56			1·00	1	00			
1	25 mm M.I. union		40		40	4	02	0·50		50	3	14	
2·44 m	32 mm medium black steel tube for open vent		30		73			0·49		49			
1	pipe hanger (purpose made)		75		75								
2	32 mm M.I. elbows		13		26			1·10	1	10			
1	80 mm × 80 mm × 32 mm M.I. tee for connecting open pipe to main		95		95	2	69	1·00	1	00	2	59	
	Extra time for high level pipework (Section 1, Chap. 4)							1·69			1	69	
Carried forward						15	34	12·94			12	94	

system in a new church hall, as illustrated in Figures 1 to 4 inclusive (Chapter 2). The job is assumed to be six miles from the office.

Quantities are first taken off from the drawings, entered on the sheets, and priced from the manufacturers' catalogues and discount sheets.

A common practice is for the estimator to compile a *materials prices book*, from which the current net prices of all equipment and materials can be instantly taken, without having to consult several catalogues and discount sheets.

The time required for erection can now be calculated by reference to the labour tables in Chapter 4.

Referring to sheet 1 of the Bill of Quantities, the labour constant for fixing the expansion tank, Section 25, Chapter 4, is 4 hours for a fitter and mate, and includes for all the work detailed in Section 25.

BILL OF QUANTITIES FOR HEATING HALL (*Contd.*)

Sheet No. 2 *Date:* 25.6.71

Main Hall		Materials						Labour					
		Unit net price						Hours fixing					
		£	p	£	p	£	p		£	p	£	p
	Brought forward					15	34	12·94			12	94
22·25 m	80 mm medium steel tube		85	20	91			9·47	9	47		
4	special pipe hangers	1	20	4	80							
2	80 mm pipe brackets		38		76							
1	80 mm pipe sleeve*		95		95							
2	80 mm M.I. bends	1	20	2	40			3·00	3	00		
1	80 mm to 65 mm twin elbow	2	25	2	25	32	07	1·50	1	50		
	Extra time for high level pipe-work							1·39	1	39	15	36
	Carried forward					47	41	28·30			28	30

BILL OF QUANTITIES FOR HEATING HALL (*Contd.*)

Sheet No. 2 (*Contd.*)

Main Hall		Materials						Labour				
		Unit net price						Hours fixing				
		£	p	£	p	£	p		£	p	£	p
	Brought forward					47	41	28·30			28	30
65 m	65 mm medium steel tube		66	42	90			22·75	22	75		
24	65 mm pipe brackets		27	6	48							
2	100 mm plain pipe sleeves*	1	01	2	02							
7	65 mm M.I. bends		96	6	72			6·30	6	30		
4	65 mm M.I. elbows		54	2	16			3·60	3	60		
4	65 mm M.I. unions		90	3	60			3·60	3	60		
1	65 mm × 25 mm × 65 mm M.I. tee		60		60	64	48	0·90	0	90	37	15
10	Hospital C.I. radiators each 4·76 m² 848 mm × 146 mm 17 section. Total heating surface 47·6 m²	3	44	163	74							
10	radiator top stays		16	1	60							
20	radiator wall brackets		16	3	20							
10	25 mm radiator wheel valves	2	40	24	00							
10	25 mm radiator L.S. valves	2	40	24	00							
1	3 way radiator valve key		35		35							
1	3 mm air valve key		10		10							
20	65 mm × 65 mm × 25 mm tongue tees		72	14	40							
12·22 m	25 mm medium steel tube		25	3	06	234	45	60·00	60	00	60	00
	Carried forward					346	34	125·45			125	45

* Unit price of sleeves includes, materials, making on site, and fixing.

BILL OF QUANTITIES FOR HEATING HALL

Sheet No. 3 *Date:* 25.6.71

Boiler room		Materials						Labour				
		Unit net price						Hours fixing				
		£	p	£	p	£	p		£	p	£	p
	Brought forward					346	34	125·45			125	45
1	Cast iron sectional boiler, heating surface 3·30 m² Output 45·78 kW	110	0	110	0							
1	safety valve	7	45	7	45							
1	thermometer and altitude gauge	4	50	4	50							
1	20 mm empty cock	2	05	2	05							
1	steel insulating jacket	15	00	15	00							
0·61 m	of 200 mm smoke pipe	6	23	3	80							
1	200 mm smoke bend	4	.85	4	85							
1	250 mm sleeve	3	15	3	15	150	80	17·00	17	00	17	00
2·13 m	80 mm medium steel tube		85	1	81			0·89	0	89		
2	80 mm flanged bends	1	80	3	60			3·00	3	00		
4	80 mm flanges and joints		55	2	20							
3·66 m	65 mm medium steel tube		66	2	42			1·28	1	28		
6	65 mm steel flanged bends	1	07	6	42			9·00	9	00		
2	65 mm steel gate valves	7	20	14	40			3·00	3	00		
12	65 mm flanges and joints		45	5	40							
2	65 mm barrel nipples		26		52							
80	22 mm × 63·5 mm M.S. bolts, nuts and washers		05	4	00							
	Hemp and lubricating oil				50	41	27					
	Add extra time for working in basement (Section 1, Chap. 4)							0·85	0	85	18	02
						538	41	160·47			160	47

Within the industry, minimum rates of pay are governed by the National Agreement mentioned earlier. For the purpose of this, and the following examples, a combined rate, for fitter and adult mate of £1·00 per hour is used, and the net labour charge for fixing the expansion tank will therefore be:

$$4 \text{ hours} \times £1·00 \text{ per hour} = £4·00$$

Sheet No. 1 gives the cost of materials and labour to complete the feed and expansion system, and is set out to enable the cost of each section of the work to be seen at a glance. Sheet No. 2 deals similarly with the main hall and Sheet No. 3 the work in the boiler room. Costs of materials and labour are entered in the first and fourth cash columns respectively, and totalled in the second and fifth cash columns of each sheet, the net cost of all materials being totalled in cash column 3, and net cost of labour in cash column 5, of the last sheet (Sheet No. 3).

SUMMARY OF TENDER
SUB-CONTRACT FOR CHURCH HALL HEATING

	£	p
(1) Total cost of materials	538	41
(2) Cost of labour, 196 hours at £1.00	196	00
(3) National Health, Redundancy Fund, Holiday Credits, Sickness and Accident Insurance	22	35
(4) Travelling allowance	18	00
(5) Selective Employment Tax	11	76
(6) Add 15 per cent on items 2, 3, 4 and 5 to allow for sub-contract conditions	37	22
(7) Cartage of materials. 4 journeys at 3 hours per journey. 12 hours at £1.40 per hour	16	80
(8) Insulating pipes in boiler room	29	65
(9) Wear and tear of tools. 3 per cent of net labour cost	5	88
	876	07
(10) Overheads and profit add 15 per cent	131	41
	1007	48
(11) P.C. cold water supply to feed and expansion tank, including $2\frac{1}{2}$ per cent for heating contractor	26	00
	1033	48
(12) Main contractors $2\frac{1}{2}$ per cent discount, 1/39th of £1 033	26	49
Total amount of estimate	1 059	97

Price to tender—£1060.

The labour for erection is estimated at 160 hours, and to this must be added the hours for filling up and testing, cleaning up after completion, and for site supervision during erection. Supervision from the office will be covered in the Establishment or Overhead Charges.

Section 34, Chapter 4, gives the time as 24+8 per cent of the hours required for erection, which in this case will be:

$$24+\left(\frac{8\times 160 \text{ hours}}{100}\right)=24+12\cdot4=36 \text{ hours.}$$

The total estimated time to complete the installation will be:

$$160+36=196 \text{ hours, for one pair of men.}$$

National Insurance, Redundancy Fund, and Holidays with Pay

The employer's contributions for National Insurance, Redundancy Fund, and the Holidays with Pay schemes must be charged to the contract, and at the tendering stage the charge can be related to the estimated hours and labour.

The employer's weekly contributions are:* National Insurance and Redundancy Fund 95p for adult males; Holidays with Pay and Sickness and Accident Premiums, craftsmen 143p and adult mate 123p, making a total weekly charge per pair (fitter and mate) of 456p.

Based on a 40 hour week the hourly rate for addition to the labour estimate to cover this cost will be:

$$\frac{456p}{40 \text{ hours}}=11\cdot40p \text{ per hour per pair of men.}$$

To provide for this in the tender now under consideration, the heating contractor will include: Estimated labour in hours × 11·40p = 196 hours × 11·40p = 2235p = £22·35.

Daily Travelling Allowance

The National Agreement provides for the payment to operatives of daily travelling allowances between home and job, according to the distance of the job from the centre, up to a maximum of 20 miles.

In the present case the distance is 6 miles, and the daily amounts payable by the employer are:

Craftsman 39p, mate 33p, which totals 72p per day.

If a five-day week is observed, the estimated number of days for which payment must be included in the tender is determined by dividing the estimated labour, in hours, by 40 and multiplying by 5.

The number of days in the present case is:

$$\frac{196}{40} \times 5 = 24 \cdot 5 \text{ (say 25 days),}$$

and the amount to include in the tender will be:

$$25 \text{ days} \times 72\text{p per day} = \pounds 18 \cdot 00$$

Selective Employment Tax*

For a fitter and mate, the hourly charge to cover Selective Employment Tax will be:

$\dfrac{2 \times 120\text{p}}{40} = 6\text{p}$, and the item to be included in this tender will amount

to $196 \text{ h} \times 6\text{p} = 1176\text{p} = \pounds 11 \cdot 76$.

Sub-contract Work

Mention was made in Chapter 1 of the extra supervision usually required on sub-contract work as compared with direct contracts. To cover the cost of this, and other hazards over which the heating or the ventilating sub-contractor has no control, such as waiting for attendance by other trades (bricklayers, joiners, plumbers, plasterers, and painters) and sometimes having to wait too long for payment by the main contractor, the labour estimate should be increased by 10 to 15 per cent, depending upon the nature and size of the installation. The application of this charge will of course be at the sub-contractor's discretion, in the light of his past experience.

Covering the Cost of the Bond

Where the value of a direct contract or sub-contract exceeds a certain sum, the heating contractor may be called upon to provide a bond of 5 per cent of the Contract Sum as security for its due performance. If the Conditions of Contract call for this security the heating contractor will include its cost in his tender.

*Based on the statutory rate at the time of writing.

ESTIMATING FOR MECHANICAL VENTILATION WORK

Sheet metal can be made in such a variety of shapes and sizes that it is impractical for manufacturers to provide catalogues and price lists covering the day-to-day requirements of the trade. Heating is somewhat different, because the distribution of water, steam, gas, and oil is generally restricted to the use of pipes in a rather limited standard range which, being freely available from stock, are more easily catalogued.

Manufacturers will quote for supply and delivery of ductwork, brackets, boltings and fixings to the site ready for trimming, jointing and erection, and the prices quoted together with the labour constants given in Section 27, Chapter 4, provide a sound basis for estimating.

In the absence of a manufacturer's quotation, an approximate estimate can be obtained by using the current *rate per tonne of ductwork* delivered to site, in conjunction with Section 27, Chapter 4.

In the example given here, a tender for the factory installation illustrated in Figure 6 (Chapter 3) is prepared on this basis, the cost of ductwork and fixings being taken as £450 per tonne, delivered to site.

The weight of ductwork and fixings has been estimated by use of tables 2, 3, 4, and 5 in Chapter 3, 10 per cent being added to allow for laps, rivets, and scrap.

Prices of fan, air heater, filter, and thermostatic steam control valve are manufacturers' net quotations. All other materials for connecting the air heater to the main steam and condensate lines are priced from makers' lists.

The time required for erection is estimated from the labour schedules in Sections 1, 27, 28, 29, 30, and 34, Chapter 4.

BILL OF QUANTITIES, FACTORY VENTILATION

Sheet No. 1   *Date :* 25.6.71

No. 1 Shop		Materials					Labour				
		Weight kg				Hours fixing					
			£	p	£	p		£	p	£	p
6	downpipes, 4·60 m × 220 mm × 24 BG	96					13·80	13	80		
6	220 mm twin bends at discharge	26					4·80	4	80		
6	220 mm bends at branch with main duct	19					4·80	4	80		
6	dampers	27					4·80	4	80		
12	25 mm × 3 mm brackets for downpipes	14									
12·2 m	220 mm × 24 BG duct at high level	42					6·10	6	10		
6	25 mm × 3 mm brackets for above	7									
12·2 m	275 mm × 24 BG duct at high level	53					7·32	7	32		
2	275 mm × 220 mm × 220 mm branch pieces	6					2·20	2	20		
6	25 mm × 3 mm brackets	7									
6·1 m	341 mm × 24 BG high level duct	33					5·19	5	19		
2	341 mm × 275 mm × 220 mm branch pieces	8					2·70	2	70	51	71
116	10 mm dia × 50 mm bolts and nuts	7									
232	10 mm washers	2									
4	25 mm × 3 mm brackets	7									
		354	159	30	159	30					
	Carried forward				159	30	51·71			51	71

BILL OF QUANTITIES, FACTORY VENTILATION (*Contd.*)

Sheet No. 2 *Date:* 25.6.71

No. 2 Shop		Materials				Labour				
	Weight kg					Hours fixing				
		£	p	£	p		£	p	£	p
Brought forward				159	30	51·71			51	71
6 downpipes, 4·6 m × 24 BG 220 mm dia × 24 BG	96					13·80	13	80		
6 220 mm bends at branches with main duct	19					4·80	4	80		
6 220 mm twin bends at discharge	26					4·80	4	80		
6 220 mm dampers	27					4·80	4	80		
12 25 mm × 3 mm flat iron brackets	14									
6·1 m 341 mm dia × 24 BG duct at high level	42					5·19	5	19		
4 25 mm × 3 mm flat iron brackets	7									
12·2 m 385 mm dia × 24 BG duct at high level	64					12·81	12	81		
2 385 mm × 341 mm × 220 mm branches	8					3·00	3	00		
6 25 mm × 3 mm flat iron brackets	10									
5·2 m 419 mm dia × 24 BG high level duct	38					5·72	5	72		
2 419 mm × 385 mm × 220 mm branch pieces	10					3·20	3	20		
	361	162	45	162	45	3·20	3	20	58	12
Carried forward				321	75	109·83			109	83

BILL OF QUANTITIES, FACTORY VENTILATION (*Contd.*)

Sheet No. 2 (*Contd.*)

No. 2 Shop		Materials					Labour				
		Weight kg					Hours fixing				
			£	p	£	p		£	p	£	p
	Brought forward				321	75	109·83			109	83
7	25 mm × 3 mm flat iron brackets	11									
15 m	440 mm dia × 24 BG high level duct	110					17·25	17	25		
2	440 mm × 419 mm × 220 mm branches	12					3·40	3	40		
2	440 mm dia × 24 BG bends	16					3·40	3	40	24	05
6	25 mm × 3 mm flat iron brackets	15									
170	10 mm dia × 50 mm bolts and nuts	9									
340	10 mm washers	3									
		176	79	20	79	20					
	Carried forward				400	95	133·88			133	88

BILL OF QUANTITIES, FACTORY VENTILATION (*Contd.*)

Sheet No. 3   *Date:* 25.6.71

Plant Room		Materials					Labour				
	Weight kg					Hours fixing					
		£	p	£	p			£	p	£	p
Brought forward				400	95	133·88				133	88
1 650 mm dia × 22 BG flanged bend at fan outlet	24					2·60		2	60		
1 650 mm dia × 1 050 mm × 1 200 mm × 20 BG flanged making up piece between fan inlet bend and air heater	54					5·20		5	20		
2 flanged make-up pieces between air heater and filter, and between filter and fresh air inlet grill. 1 200 mm × 1 050 mm × 20 BG, each 310 mm long	73					2·80		2	80		
1 440 mm × 440 mm × 650 mm × 22 BG twin bend at fan outlet, flanged	27					1·70		1	70	14	00
1 m make-up piece between the above, and the fan outlet, flanged	18					1·70		1	70		
2 flanges for above	7										
2 40 mm × 4 mm flat iron straps for above	10										
200 12·5 mm × 50 mm bolts and nuts	23										
400 12·5 mm washers	5										
	241	108	45	108	45						
Carried forward				509	40	147·88				147	88

BILL OF QUANTITIES, FACTORY VENTILATION (*Contd.*)
Sheet No. 3 (*Contd.*)   *Date:* 25.6.71

Plant Room	Materials					Labour				
	Weight kg					Hours fixing				
		£	p	£	p		£	p	£	p
Brought forward				509	40	147·88			147	88
1 slow speed multi-vane centrifugal fan rated at 3·54 m/second, at 3·11 mbar, with 650 mm dia inlet and 700 mm × 500 mm outlet	204	00				16·00	16	00		
1 set anti-vibration mountings and foundation bolts	25	00								
1 1 050 mm × 1 200 mm high single row copper gilled tube air heater, with 32 mm steam and 25 mm conden-sate tappings	85	00				12·50	12	50		
1 1 050 mm × 1 200 mm air filter 'throw-away' wool type	171	60				6·00	6	00	34	50
1 1 050 mm × 1 200 mm louvred timber inlet grill, to be provided and fixed in outside wall by builder										
2 canvas connec-tions to fan	6	20	491	80						
Carried forward				1 001	20	182·38			182	38

BILL OF QUANTITIES, FACTORY VENTILATION (*Contd.*)

Sheet No. 4   *Date:* 25.6.71

	Plant Room	Unit net cost £	p	£	p	£	p	Hours fixing	£	p	£	p
	Brought forward					1 001	20	182·38			182	38
1	thermostat in fan discharge duct	12	50	12	50			2·00	2	00		
1	32 mm thermostatic steam control valve on heater with 2·44 m of capillary tube to thermostat	42	00	42	00			1·20	1	20		
3 m	32 mm steam pipe		27		81			0·60		60		
2	32 mm pipe brackets		14		28							
1	32 mm steam stop valve	4	25	4	25			0·55		55		
1	100 mm steam gauge	2	00	2	00			0·50		50		
1	15 mm automatic air vent	5	10	5	10			0·35		35		
1	80 mm × 80 mm × 32 mm tee in steam main	1	65	1	65			1·00	1	00		
2	32 mm bends		29		58			1·10	1	10		
1	32 mm union		80		80			0·55		55		
6 m	25 mm galvanised condensate pipe		23	1	38			1·14	1	14		
4	25 mm pipe brackets		10		40							
2	25 mm nipples		5		10							
2	25 mm unions		60	1	20			1·00	1	00		
8	25 mm bends		23	1	84			4·00	4	00		
1	50 mm × 50 mm × 25 mm tee in condensate main		90		90			0·80		80		
1	25 mm steam trap and strainer	13	00	13	00			1·00	1	00		
1	back pressure valve	2	30	2	30			0·50		50		
2	25 mm tees for by-pass		32		64			1·00	1	00		
3	25 mm steam valves for steam trap isolation and by-pass	2	50	7	50	99	23	1·50	1	50	18	79
	Carried forward					1 100	43	201·17			201	17

BILL OF QUANTITIES, FACTORY VENTILATION (*Contd.*)

Sheet No. 5   *Date:* 25.6.71

Plant Room		Materials					Labour				
	Unit net cost						Hours fixing				
	£	p	£	p	£	p		£	p	£	p
Brought forward					1 100	43	201·17			201	17
jointing and sundry materials			3	10							
3 m 32 mm pipe and fittings insulated to standard specification. Section (d2) of Clause 38			4	80							
6·1 m 25 mm pipe and fittings insulated to standard specification. Section (d2) of Clause 38			9	35							
1 triple pole galvanised switch-fuse to control fan circuit			4	70			1·50	1	50		
1 three phase fan starter			13	00	34	95	1·50	1	50	3	00
Extra labour for fixing plant on raised platform, 15% of 34·5 hours							5·18	5	18		
Extra labour for erecting pipework on raised platform 10% of 18·79 hours							1·88	1	88	7	06
					1 135	38	211·23			211	23

SUMMARY OF TENDER
SUB-CONTRACT FOR FACTORY VENTILATION

	£	p
(1) Total cost of materials	1 135	38
(2) Labour for erection, 211 hours at £1.00	211	00
(3) Labour, testing, clearing site, and supervision on site: 24+8% of item (2)=41 hours	41	00
(4) National Insurance, Redundancy Fund, Holiday Credits, Sickness and Accident Insurance: 252 hours at 11·40 p	28	74
(5) Travelling Allowances (job 14 miles from centre) 32 days at 100p	32	00
(6) Selective Employment Tax. 252 hours at 6p	15	12
(7) Add 15% on items 2, 3, 4, 5 and 6 to allow for sub-contract conditions	49	18
(8) Cartage of heating materials. (Ventilation equipment delivered to site by manufacturer) 4 hours at £1.40	5	60
(9) Wear and tear of tools: 3% on net labour costs of £252	7	56
	1 525	58
(10) Overheads and profit, add 15%	228	84
	1 754	42
(11) P.C. for electrical wiring, including 2½% for ventilating contractor	43	86
Provisional sum for contingencies	100	00
	1 898	28
(12) Main contractors 2½% discount, 1/39th of £1 898	48	66
Total amount of estimate	1 946	94

Price to tender—£1 947

THE PREPARATION OF PRELIMINARY ESTIMATES

The Prices Record Book

Preliminary estimates for heating and/or ventilating are generally wanted months before the architect is in a position to provide the engineer with finished plans and elevations. The owner may be a Local Authority or commercial organisation, who must be advised of the probable cost of a project at the early planning stage. This budgeting information is usually needed by the committee or board of directors responsible for initiating and approving the new building project.

In these circumstances the engineer will be asked to give an estimate based upon sketch plans. Such estimates, while not considered firm, are expected to be reasonably close to the ultimate cost of the work. A quick estimate can be obtained by comparing the job in hand with similar jobs with which the office has been concerned in the not too distant past.

Cost particulars of all work passing through the office are entered briefly in a price record book. The consulting engineer's records are built up from tender results, and in the case of contractors from the actual final costings of completed contracts. Normal costing methods will give the contractor sufficient up-to-date information to enable the probable cost of future work to be approximately estimated. From his cost books the contractor may build up schedules of the actual cost of different parts of schemes he has installed in various types of buildings, such as boiler plant, hot water cylinders, pumps, automatic stokers, oil burners, radiators, pipework, etc. When asked to submit a quick approximate price for new work, the proposed scheme is compared with the specification and price of a completed installation of a similar nature.

The heat requirements of the new scheme are compared with those of the completed building in order to arrive at a proportionate figure for the new scheme. An example will make the method of estimating clear. Suppose an estimate is required for a school having a heat loss (calculated from the sketch plans) of 117 kW (400 000 Btu/h) and needing 40 hot water points. From the price record book cost particulars of a similar school completed twelve months previously are given as:

Heat loss	103 kW (350 000 Btu/h)
Hot water	34 points
Heating. Final cost as installed	£7 000
Hot water system as installed	£2 380

$$\text{Heating cost per kW} = \frac{£7\,000}{103} = £68.$$

$$\text{Hot water, cost per point} = \frac{£2\,380}{34} = £70.$$

Two other factors must be taken into account in determining the probable cost of the proposed scheme. These are:

(1) Any advance in wages and/or the cost of materials which has occurred since the comparable installation was completed.
(2) The distance of the proposed work from the contractor's office.

The site of the contract completed a year previously may be some miles nearer to the contractor's headquarters than the present site, in which case a suitable increase, to cover travelling time and allowances, and extra cartage costs becomes necessary.

During the interval of twelve months, labour and materials costs may have risen and should be accounted for in the new estimate.

The approximate estimated cost of the proposed scheme will be:

$$
\begin{array}{ll}
\text{Heating. 117 kW at £68 per kW} & \\
& = 117 \times £68 \\
& = £7\,956 \\
\text{H.W.S. 40 points at £70} & = £2\,800 \\
\hline
& £10\,756 \\
\begin{array}{l}\text{Add 10 per cent for extra}\\ \text{distance and rise in}\\ \text{labour and}\\ \text{materials costs}\end{array} & = £1\,076 \\
\hline
\text{Total} \quad £11\,832 \\
\hline
\end{array}
$$

The figure should be rounded off to the nearest hundred pounds, and in this case the contractor would submit a preliminary estimate of £11 800.

The heating consultant, whether in private practice or employed in the Architect's Department of the Local Authority or business organisation, will have direct knowledge of tender results for the classes of building with which he normally deals and this information will provide the basis for compiling the Price Record Book.

From the tender results of schemes of his own design, the consultant's records are built up as shown in Figure 14, which represents a specimen price record sheet.

When inviting tenders the engineer will require each contractor to split up his tender to show clearly the amount allowed for each main section of the work, viz: space heating, hot water supply, gas services, and any other additional services which may be included, such as ventilating fans, cold water supply, etc.

The first five columns are entered up during the design stage and the remaining columns after tenders have been received.

The site location, which has a distinct bearing on cost, is noted in Column 12.

When estimating from the Price Record Book the index building chosen must have similar usage; be of the same type; have similar heating plant and a heat loss reasonably near to that of the proposed new building.

Example. Give a preliminary estimate of the cost of heating and hot water supply for a school to be erected in an industrial area, in which several good heating contractors are established.

Heat loss of new school as calculated from architect's sketch plans   114 kW (390 000 Btu/h)

Boiler plant   Sectional C.I. boilers with automatic underfeed coal stokers.

Number of H.W.S. points   36

School C, Figure 14, is chosen as nearly fulfilling the requirements of the new project, and the probable cost can be estimated in the following manner:

> *Heating*
> Heat loss of new school × School C, Col. 9
> = 114 × £61
> = £6 945
>
> *H.W.S.*
> Number of points in new school × School C, Col. 10
> = 36 × £63
> = £2 268

Summary

Heating	£6 945
H.W.S.	£2 268
	£9 213
Add for rise in costs since date of School C tenders 10 per cent	£921
Preliminary estimate	£10 134
Say	£10 100

Approximate estimating for ventilation work is also simplified when price records of past jobs are available.

Figure 15 is a specimen page suitable for a ventilation price book in which all the cost information necessary for estimating by comparison can be quickly recorded.

The rules regarding similarity of building; building usage; size of load; and site location characteristics applicable to heating estimates must also be observed when estimating for ventilation.

(1) Job title Date	(2) Boiler plant	(3) Building heat loss kW	(4) No. of H.W.S. points	(5) No. of gas points	(6) Average heating tender £	(7) Average H.W.S. tender £	(8) Average gas tender £	(9) Heating cost per kW £	(10) H.W.S. cost per point £	(11) Gas cost per point £	(12) Special features of the scheme which affect the total cost
School A Sept., 1970	Coke hand-fired	33·12	Nil	Nil	1250	Nil	Nil	38	Nil	Nil	Location 30 miles from city
School B Dec., 1970	Auto. stoker coal	92·32	31	8	6300	2160	194	68	70	24	School situated in country district remote from large town
School C June, 1970	Auto. stoker coal	97·29	29	7	5900	1840	320	61	63	46	School situated near city, and close to heating contractor's offices
Divisional Police Station Dec., 1970	Auto. stoker coal	69·16	15	Nil	3600	540	Nil	53	36	Nil	Cold water included in mechanical engineering sub-contract. Average tender for 30 points: £420=£14 per point. H.W.S. by individual electric heaters
Girls' Hostel June, 1971	Coke hand-fired	53·93	44	Nil	2700	2710	Nil	50	62	Nil	H.W.S. plant and heating mains installed large enough to meet needs of future extension. Location—large town
Old Folks Hostel A Oct., 1970	Auto. stoker coal	122·6	87	5	6650	4000	143	54	46	29	Job within 5 miles of large town and within easy reach of heating contractor's office
Old Folks Hostel B April, 1971	Auto. stoker coal	107·1	85	5	6900	3720	118	64	44	24	Job within 8 miles of city and within easy reach of heating contractor's office

Figure 14. Specimen page of Price Record Book (Heating and H.W.S.)

(1)	(2)	(3)	(4)	(5)	(6)	(7)	(8)	(9)	(10)	(11)
Job title and date	Fan size m³/sec total head	Fan control type	Cost of fan as installed and connected to ducts £	Cost as Col. 4 + filter £	Cost as Col. 4 + heater £	Cost as Col. 4 + heater + filter £	Cost as Col. 4 + heater + washer £	Weight of sheet metal ducts tonnes	Cost per tonne of ductwork supplied and fixed £	Special features of the installation which affect the total cost

Figure 15. Specimen page of Price Record Book (Ventilation)

ESTIMATING FOR JOBBING WORK

Firm prices for repairs and replacements can only be given when the work to be done is capable of being accurately measured. Jobbing work falls into two distinct categories:

(1) Work which can be seen and measured beforehand.
(2) Work which cannot be measured, or is difficult to measure beforehand and must therefore be carried out on a time and material, or daywork basis.

Work in category (2) should always be accepted on a daywork basis.

In a highly specialised craft such as space heating, water heating, and ventilation, the contractor is usually called in to diagnose the trouble and this may well take much longer than the actual repair.

The costing of jobbing work is dealt with in Chapter 6, and an example given.

Generally speaking category (1) embraces work on boilers, cylinders, pumps, automatic stokers, oil burners, fans, air filters, air heaters, and their component parts.

For the normal run of this class of work the labour constants given in Chapter 4 apply, and the quotation can be based upon a brief Bill of Quantities. When the enquiry involves the repair or replacement of specialist's equipment, such as automatic stokers or oil burners, the contractor will base his price upon the manufacturer's quotation to him for doing the work, plus his own charge for supervision, establishment costs, and profit.

Example 1. Estimate the price to quote for replacing a 200s oil burner unit serving a hot water heating boiler rated at 110·8 kW (378 000 Btu/h).

	£
Manufacturer's quotation for dismantling and taking away old oil burner; re-brick boiler, provide and fix new oil burner, re-connect existing wiring and controls, and test	270·00
Heating contractor's charge to cover preliminary inspection, supervision of work, and establishment charges, 15 per cent	40·50
	310·50
Contractor's profit, 10 per cent	31·05
Price to quote	341·55

*Example 2. Estimate the price to quote for dismantling and removing
from site one 52·76 kW (180000 Btu/h) H.W.S. boiler and primary
pipework, and renewing completely. Supply and fit new mountings,
and test. The work to be carried out during the weekend. Site location,
11·26 km (7 miles) from office. (Basement boiler room.)*

A list of quantities is prepared on the site, and using Sections 1,
17, 19, 34, and 35, Chapter 4, a priced Bill of Quantities is completed.
Materials are priced from the manufacturers' net trade price lists.

Insulation of pipes is quoted at current rates inclusive of labour.

Labour for dismantling is calculated from Sections 1, 17, and 19,
Chapter 4, multiplied by the constants given in Section 35.

Filling up, testing, cleaning up, and supervision, is calculated from
Section 34, Chapter 4.

Payment for overtime, travelling allowances, and dirty money for
dismantling, will be made in accordance with the terms of the
National Agreement current at the date of tendering.

The work will be carried out during the weekend, starting at
8 a.m. on Saturday and finishing on Sunday, the overtime rate being
time and a half for the first four hours on Saturday, and double time
from then until the job is completed on Sunday night.

**BILL OF QUANTITIES FOR RENEWAL OF H.W.S. BOILER AND
PRIMARY CIRCULATION**

Sheet No. 1   *Date:* 25.6.71

Work in boiler room in basement of occupied building	Materials						Labour			
	Unit net price						Hours fixing			
	£	p	£	p	£	p	h	h	£	p
Dismantle, load on lorry and remove H.W.S. boiler. Heating surface 2·6 m² Dismantle, load on lorry and remove 6 m, 80 mm and 4 m, 80 mm flanged bends, and 2 m, 203 mm smoke pipe and fittings							8·00 4·89	12·89	12	89
Carried forward							12·89		12	89

Sheet No. 1 (*Contd.*)

BILL OF QUANTITIES FOR RENEWAL OF H.W.S. BOILER AND PRIMARY CIRCULATION

Work in boiler room in basement of occupied building		*Materials*						*Labour*			
		Unit net price						*Hours fixing*			
		£	p	£	p	£	p	h	h	£	p
	Brought forward							12·89		12	89
	Provide and fix 1 new sectional H.W.S. boiler, of 2·6 m² heating surface. Rated output 52·76 kW, continuous operation	225	00	225	00						
1	20 mm spring loaded safety valve	6	60	6	60						
1	combined altitude gauge and thermometer	4	25	4	25						
1	20 mm empty cock with hose connection	2	70	2	70						
1	galvanised steel boiler jacket	22	00	22	00	260	55	16·00			
2 m	203 mm dia steel smoke pipe	4	06	8	12						
1	203 mm dia steel smoke bend	6	40	6	40	14	52	2·50	18·50	18	50
6 m	80 mm galvanised steel tube		88	5	28			2·52			
2	80 mm pipe hangers		85	1	70						
4	80 mm flanged steel bends	1	86	7	44			6·00	8·52	8	52
8	80 mm brass joint rings		23	1	84						
32	16 mm × 65 mm nuts, bolts and washers		05	1	60						
	Joint compound, hemp and lubricating oil		50		50	18	36				
	Add 10% to boiler labour (occupied building)							1·60			
	Add 25% to pipe-work labour (basement in occupied building)							2·13	3·73·	3	73
						293	43		43·64	43	64

SUMMARY OF TENDER
RENEWAL OF H.W.S. BOILER

	£	p
(1) Total cost of materials	293	43
(2) Labour for dismantling and erection. 44 hours at £1.00	44	00
(3) Labour for site supervision, filling up, testing and clearing away after completion. 30 per cent of item (2)	13	20
(4) Overtime. Two pairs of men working from 8 a.m. Saturday to 9-30 p.m.; 8 a.m. Sunday until completion of job. At National Agreement overtime rates	39	00
(5) Abnormal conditions money. Dismantling old sectional boiler, smoke pipe, and old primary pipework (12·89 hours) say 6·50 h for two pairs. 28p per day × 4	1	12
(6) Travelling Allowance. 2 days × 2 pairs × 72p	2	88
(7) Wear and tear of tools: 3 per cent of item (2)	1	32
(8) Cartage of materials. Old boiler and pipework from site. New pipework and tools to and from site. (New boiler delivered to site by manufacturer). Three journeys. 9 h at £1.40p	12	60
(9) Insulation to primary pipework with 19 mm thick fibre glass	10	94
	418	49
(10) Overheads and profit, 20 per cent	83	69
Total amount of estimate	502	18

Price to tender—£502.

The amount to include for Establishment Charges is usually higher for repair work than for new work, to cover the cost of the initial inspection, measuring up and estimating, and also for the frequent visits by technical office staff while the work is in progress. Compared with new contracting work, it is usual to increase this charge to not less than $1\frac{1}{2}$ times that for new work.

When work of a maintenance nature has to be carried out during the weekend, and completely outside the normal five-day week, no charge is made for National Insurance and Holidays with Pay contributions, as these will usually be charged to work upon which the men are engaged during the normal working week, generally Monday to Friday inclusive.

Labour employed on Jobbing Work
Labour employed on jobbing work executed on a time and materials basis is usually charged at an all-in hourly rate, which also covers National Insurance, Selective Employment Tax, Redundancy Fund, Holidays with Pay, wear and tear of hand tools, and Establishment Charges.

The heating contractor will decide the rate to charge for day-to-day jobbing work embracing the usual run of repairs and replacements of existing equipment and/or pipework made necessary by the normal process of wear and tear, accident, additions, minor extensions and servicing.

For maintenance and service work of a regular nature, e.g. for public buildings, business properties, factories, property companies, and housing estates, the heating contractor and client may negotiate competitive rates. In such negotiations, the contractor must take into account his own trading position, the general nature of the work, the client's trading and financial position, and his reputation regarding prompt settlement of accounts.

Regarding normal day-to-day jobbing work mentioned above the following rates will produce a reasonable return.

(1) *Labour*. Plus 60 per cent. To cover employer's liability and third party insurances, National Health and Unemployment Insurance, Redundancy, S.E.T., Holidays with Pay, Sickness Insurance, use of hand tools, and general establishment charges.

(2) *Overtime*. According to National Agreement.

(3) Out Working Allowances, fares, abnormal conditions, radius allowance. Plus 20 per cent.

(4) Materials, carriage, and cartage. Plus 20 per cent.

(5) Sub-contracts. Plus 20 per cent.

For jobbing work carried out by the heating contractor as an indirect contract, the main contractor's cash discount, usually $2\frac{1}{2}$ per cent, must be provided by the addition of one thirty-ninth of the total value of the work.

PREPARING A TENDER FOR SMALL BORE DOMESTIC CENTRAL HEATING AND HOT WATER SUPPLY

To be installed in a Three Bedroom Semi-detached House

The bill (sheets 1 to 6), covers the provision and installation of a two-pipe Small Bore Central Heating and Hot Water System in the three bedroom semi-detached house, Figure 16.

Each section of the work is costed separately, commencing with the expansion tank arrangement in the roof space, and the total costs of materials, equipment, and labour for the combined installation are given at the foot of Sheet 6.

Materials, and equipment, are estimated at £314·33 and labour, 117·58 hours, say 118 hours for a fitter and mate.

After erection the systems are filled up, tested and cleaned up prior to handing over, and a labour charge is necessary to cover this work, together with supervision, and time in giving operating tuition to the householder.

For commercial work on building sites where other trades are employed, this item is calculated in accordance with Section 34, Chapter 4. In the case of small domestic installations, the following allowances at the fitter and mate rate should be included in the case of occupied houses:

Filling up, testing, and cleaning up—8 hours. Supervision, and instruction to householder—10 hours, a total of 18 hours.

The total estimated time for completing the installation, will, in this case be:

$$118 + 18 = 138 \text{ hours for one pair of men}$$

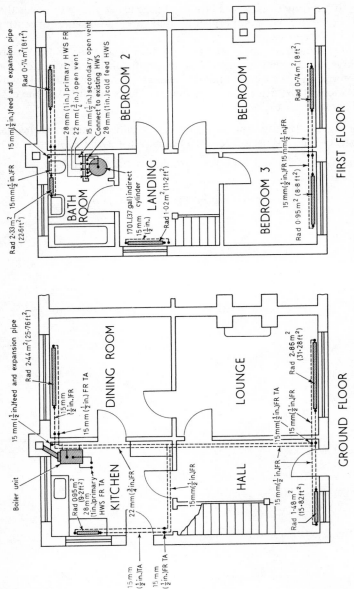

Figure 16. *Two-pipe small bore domestic central heating and hot water system in three-bedroom semi-detached house*

BILL OF QUANTITIES FOR SMALL BORE CENTRAL HEATING AND HOT WATER SUPPLY IN THREE BEDROOM OCCUPIED HOUSE. (COPPER WITH CAPILLARY JOINTS AND FITTINGS)

Sheet No. 1 Date: 25.6.71

	Work in roof space (surface pipe work)	Unit net price						Hours fixing			
		£	p	£	p	£	p	h	h	£	p
1	45 litre (10 gal) galvanised feed and expansion tank with cover (insulated)	2	95	2	95						
1	15 mm ball valve	1	45	1	45	4	40	3·00			
	Add 20% for work in occupied house							0·60	3·60	3	60
1	15 mm stop cock	1	10	1	10			0·30			
2	15 mm tank connectors		35		70			0·60			
1	28 mm tank connector (overflow)		70		70			0·35			
5 m	15 mm copper tube, for open vent F and E pipe, and C.W. connection		33	1	65			0·55			
2·2 m	28 mm copper tube for overflow pipe (B.S.2871: Part 1, Table X)		76	1	67			0·33			
1	15 mm tee		15		15			0·30			
5	15 mm elbows		08		40			1·50			
2	28 mm elbows		20		40			0·70			
4	15 mm screw-on pipe brackets		06		24			0·32			
2	28 mm screw-on pipe brackets		09		18	7	13	0·16			
	Provide and fix to 15 mm pipes:—										
5 m	19 mm thick mineral wool rigid sectional insulation		35	1	75	1	75	0·75	5·86	5	86
	Add to pipework labour, 10% for work in roof space, plus 20% for work in occupied house								1·76	1	76
	Carried forward					13	28		11·22	11	22

BILL OF QUANTITIES FOR SMALL BORE CENTRAL HEATING AND HOT WATER SUPPLY IN THREE BEDROOM OCCUPIED HOUSE
(Contd.)

Sheet No. 2   *Date:* 25.6.71

Radiators and surface space heating pipework. *First floor*	Materials						Labour			
	Unit net price						Hours fixing			
	£	p	£	p	£	p	h	h	£	p
Brought forward					13	28		11·22	11	22
First floor. Steel radiators:										
Bathroom Double panel 2·33 m²	8	10	8	10						
Bedroom (1) Single panel 0·74 m²	2	45	2	45						
Bedroom (2) Single panel 0·74 m²	2	45	2	45						
Bedroom (3) Single panel 0·95 m²	3	20	3	20						
Landing Single panel 1·62 m²	5	25	5	25						
5 radiator air valves		10		50						
10 radiator wall brackets		10	1	00						
5 15 mm radiator wheel valves	1	20	6	00						
5 15 mm radiator lock shield valves	1	20	6	00	34	95	15·75			
2 m 15 mm copper tube radiator connectors		33		66			0·22			
2·7 m 15 mm copper tube radiator connectors FE pipe		33		89			0·30			
1 15 mm build-in pipe bracket		14		14	1	69	0·25	16·52	16	52
Add 20% and 5% for occupied house and first floor work respectively, to labour								4·13	4	13
Carried forward					49	92		31·87	31	87

BILL OF QUANTITIES FOR SMALL BORE CENTRAL HEATING AND HOT WATER SUPPLY IN THREE BEDROOM OCCUPIED HOUSE
(*Contd.*)

Sheet No. 3   *Date:* 25.6.71

Radiator and surface space heating pipework. Ground floor		Materials					Labour		
		Unit net price					Hours fixing		
		£	p	£	p	£	p		
		£	p	£	p	£	p	h	h	£	p

	Radiator and surface space heating pipework. Ground floor	£	p	£	p	£	p	h	h	£	p
	Brought forward					49	92		31·87	31	87
	Ground floor. Steel radiators:										
	Dining room										
	Single panel 2·44 m²	8	15	8	15						
	Lounge										
	Single panel 2·86 m²	9	25	9	25						
	Hall										
	Single panel 1·48 m²	4	90	4	90						
	Kitchen										
	Single panel 0·95 m²	3	20	3	20						
4	radiator air valves		10		40						
10	radiator wall brackets		10	1	00						
4	15 mm radiator wheel valves	1	20	4	80						
4	15 mm radiator lock shield valves	1	20	4	80	36	50	11·00			
1·5 m	15 mm copper tube for radiator connections		33		50			0·17			
3 m	15 mm copper tube for FE pipe		33		99			0·33			
1	15 mm elbow for FE pipe		08		08			0·30			
7	15 mm build-in pipe brackets		14		98	2	55	1·75			
17 m	15 mm tube. Risers to first floor		33	5	61	5	61	1·87			
	Add to labour for one extra bracket to Dining Room and Lounge radiators							0·70	16·12	16	12
	Add to labour, 20% for work in occupied house								3·22	3	22
	Carried forward					94	58		51·21	51	21

BILL OF QUANTITIES FOR SMALL BORE CENTRAL HEATING AND HOT WATER SUPPLY IN THREE BEDROOM OCCUPIED HOUSE
(Contd.)

Sheet No. 4   Date: 25.6.71

| | | Materials | | | | | | Labour | | | |
	Under floor pipework on first and ground floors	Unit net price						Hours fixing			
		£	p	£	p	£	p	h	h	£	p
	Brought forward					94	58	51·21		51	21
	First floor										
16 m	15 mm copper tube (B.S.2871: Part 1, Table X)		33	5	28			1·76			
4	15 mm tees		15		60			1·20			
16	15 mm elbows		08	1	28			4·80			
9	15 mm screw-on pipe brackets		06		54	7	70	0·72	8·48	8	48
	Add to labour:										
	5% for work on first floor							0·42			
	30% for work under floors							2·54			
	20% for work in occupied house							1·70	4·66	4	66
	Ground floor										
7 m	22 mm copper tube (B.S.2871: Part 1 Table X)		55	3	85			0·91			
32 m	15 mm copper tube (B.S.2871: Part 1 Table X)		33	10	56			3·52			
4	22 mm screw-on pipe brackets		07		28			0·32			
10	15 mm screw-on pipe brackets		06		60			0·80			
2	22 mm tees		21		42			0·60			
1	22 mm × 15 mm × 22 mm tees		52	1	04			0·60			
2	22 mm × 15 mm × 15 mm tees		52	1	04			0·60			
7	15 mm tees		15	1	05			2·10			
2	22 mm stop valves	1	60	3	20			0·60			
2	15 mm empty cocks	1	45	2	90			0·60			
2	15 mm regulating valves	1	25	2	50			0·60			
17	15 mm elbows		08	1	36	28	80	5·10	16·35	16	35
	Add to labour: 30% for working under floors, and 20% for occupied house								8·18	88	18
	Carried forward					131	08		88·88	88	88

BILL OF QUANTITIES FOR SMALL BORE CENTRAL HEATING AND HOT WATER SUPPLY IN THREE BEDROOM OCCUPIED HOUSE
(*Contd.*)

Sheet No. 5   *Date: 25.6.71*

	Hot water supply system Work on first and ground floors. Surface work	Materials						Labour			
		Unit net price						Hours fixing			
		£	p	£	p	£	p	h	h	£	p
	Brought forward					131	08		88·88	88	88
	First floor										
	Dismantle and remove from site one 127 litre (28 gal) direct cylinder, and existing 20 mm (¾ in) steel primary circuit							2·40	2·40	2	40
	Disconnect 25 mm secondary and cold feed pipes and re-connect to new cylinder										
1	145 litre (32 gal) copper indirect cylinder with insulating jacket	33	45	33	45			4·00	4·00	4	00
4·5 m	28 mm copper tube (B.S.2871)		76	3	42			0·68			
6	28 mm elbows		20	1	20			2·10			
2	28 mm union adaptors		56	1	12			0·70			
2	28 mm unions		56	1	12			0·70			
1	19 mm union adaptor		41		41			0·30			
1	28 mm × 28 mm × 22 mm tee		52		52	41	24	0·35	4·83	4	83
	Add 25% to labour for first floor and occupied building								2·81	2	81
	Ground floor										
5·5 m	28 mm copper tube (B.S.2871)		76	4	18			0·83			
2	28 mm union boiler connectors		56	1	12			0·70			
5	28 mm elbows		20	1	00			1·75			
2	screw-on 28 mm pipe brackets		09		18			0·16			
	Sectional insulation. 5 m. 28 mm pipe		28	1	40	7	88	0·75	4·19	4	19
	Add 20% to labour for occupied house								0·84	0	84
	Carried forward					180	10		107·95	107	95

BILL OF QUANTITIES FOR SMALL BORE CENTRAL HEATING AND HOT WATER SUPPLY IN THREE BEDROOM OCCUPIED HOUSE
(Contd.)

Sheet No. 6   *Date:* 25.6.71

Heating and hot water supply. Small bore boiler. Provide and fix. Surface work. Ground floor		Materials					Labour			
	Unit net price						Hours fixing			
	£	p	£	p	£	p	h	h	£	p
Brought forward					180	10		107·95	107	95
1 Natural gas fired small bore boiler unit, with pump, electric heating controls and programmer, and gas safety controls. Connect to existing brick flue, and to gas and electric supplies	120	00	120	00			5·25			
1 safety valve	4	60	4	60			0·35			
1 combined altitude gauge and thermometer	4	10	4	10	128	70	0·35	5·95	5	95
2 22 mm unions		41		82			0·60			
4 22 mm elbows		11		44			1·20			
2 22 mm × 19 mm × 15 mm tees		32		64			0·60			
2 22 mm floor plates		15		30						
1·5 m 22 mm copper tube (B.S. 2871: Part 1)		55		83			0·20	2·60	2	60
Solder, paste and sundries			2	50	5	53				
Add 10% to boiler labour								0·60		60
and 20% to pipework labour for occupied house								0·48		48
					314	33		117·58	117	58

SUMMARY OF TENDER

CONTRACT FOR SMALL BORE CENTRAL HEATING AND CONVERSION OF EXISTING DIRECT HOT WATER SYSTEM TO INDIRECT OPERATION INCLUDING THE PROVISION OF A NEW INDIRECT CYLINDER

	£	p
(1) Total cost of materials	314	33
(2) Total cost of labour for erection 118 h at £1	118	00
(3) Supervision, testing, cleaning up, and operating instruction to householder: 18 h at £1	18	00
(4) Travelling allowance: 17 days at 28p	4	76
(5) National Insurance, Redundancy Fund, Holiday Credits, Sickness and Accident Insurance: 136 h at 11·40p	15	50
(6) Selective Employment Tax: 136 h at 6p	8	16
(7) Cartage of materials: 3 h at £1.40	4	20
(8) Wear and tear of tools: 3 per cent on net erection labour cost	3	54
	486	49
(9) Profit and overheads: add 15 per cent	72	97
(10) P.C. for electrical work on boiler controls. Includes 2½ per cent for heating contractor	9	00
Total amount of estimate	568	46

Price to tender—£568.

Note: The heat loss from 15 mm pipe fixed under the floors is not excessive, and in most cases is uninsulated. Where such floors are well ventilated, e.g. with air bricks on more than one side, insulation with sectional flexible foam, with split sides jointed with adhesive, is advisable at an extra cost for this three bedroom house of £23, bringing the amended tender to £591.

PREPARING A TENDER FOR MICRO BORE DOMESTIC CENTRAL HEATING AND HOT WATER SUPPLY

To be installed in a three bedroom semi-detached house.

The bill (Sheets 1 to 5) covers the provision and installation of a Micro Bore Central Heating and Hot Water Supply System in the three bedroom house (*Figure 17*).

Materials and equipment are estimated at £283·28, and labour for erection 90·74 hours, say 91 hours.

BILL OF QUANTITIES FOR MICRO BORE CENTRAL HEATING AND HOT WATER SUPPLY IN THREE BEDROOM OCCUPIED HOUSE (NOTE. ALL MICRO BORE CIRCUITS AND RADIATOR CONNECTIONS TO B.S. 2871: PART 2)

Sheet No. 1   *Date:* 25.6.71

	Work in roof space (surface pipe work)	Unit net price £	p	£	p	£	p	Hours fixing h	h	£	p
				Materials					*Labour*		
1	45 litre (10 gal) galvanised feed and expansion tank, with cover and insulation	2	95	2	95						
1	15 mm ball valve	1	45	1	45	4	40	3·00			
	Add 20% to labour for occupied house							0·60	3·60	3	60
1	15 mm stop cock	1	10	1	10			0·30			
2	15 mm tank connectors		35		70			0·60			
1	28 mm tank connector		64		64			0·35			
5 m	15 mm (B.S.2871: Part 1, Table X) copper tube for open vent, F and E pipe and C.W. service connection		33	1	65			0·55			
2·2 m	28 mm copper tube for overflow pipe		76	1	67			0·33			
1	15 mm tee		15		15			0·30			
5	15 mm elbows		08		40			1·50			
2	28 mm elbows		20		40			0·70			
4	15 mm screw-on pipe brackets		06		24			0·32			
2	28 mm screw-on pipe brackets		09		18	7	13	0·16			
	Provide and fix to 15 mm pipework: 5 m of 19 mm thick mineral wool rigid sectional insulation		35	1	75	1	75	0·75			
	Add to pipework labour, 10% for work in roof space plus 20% for work in occupied house							1·76	7·62	7	62
	Carried forward					13	28		11·22	11	22

BILL OF QUANTITIES FOR MICRO BORE CENTRAL HEATING AND HOT WATER SUPPLY IN THREE BEDROOM OCCUPIED HOUSE (Contd.)
Sheet No. 2 *Date: 25.6.71*

Work on first floor (*surface pipework and radiators*)	Materials Unit net price £	p	£	p	£	p	Labour Hours fixing h	h	£	p
Brought forward					13	28		11·22	11	22
Steel radiators										
Bathroom										
Double panel 2·33 m²	8	10	8	10						
Bedroom (1)										
Single panel 0·74 m²	2	45	2	45						
Bedroom (2)										
Single panel 0·74 m²	2	45	2	45						
Bedroom (3)										
Single panel 0·95 m²	3	20	3	20						
Landing										
Single panel 1·62 m²	5	25	5	25						
5 radiator air valves		10		50						
10 radiator wall brackets		10	1	00						
2·5 m 8 mm tube for radiator valve distribution pipes (B.S. 2871: Part 2)		21		53						
5 twin radiator valves (regulating type)	1	66	8	30	31	78	10·00	10·00	10	00
Provide and fit to existing direct cylinder, 1 hot water primary immersion heater with electric thermostatic control unit	17	00	17	00			3·00			
9·5 m 15 mm copper tube		33	3	14			1·05			
5 15 mm elbows		08		40			1·50			
1 15 mm tee		15		15			0·30	5·85	5	85
2 15 mm celing plates		13		26						
Solder, paste, sundries			1	00	21	95				
Add 5% for work on first floor plus 20% for work in occupied house							3·96	3·96	3	96
Carried forward					67	04		31·03	31	03

BILL OF QUANTITIES FOR MICRO BORE CENTRAL HEATING AND HOT WATER SUPPLY IN THREE BEDROOM OCCUPIED HOUSE (Contd.)

Sheet No. 3 *Date:* 25.6.71

	Work on first floor (under floor pipework)	Materials						Labour			
		Unit net price						Hours fixing			
		£	p	£	p	£	p	h	h	£	p
	Brought forward					67	04		31·03	31	03
1	Microbore distribution manifold 22 mm × 10–8 mm branches	5	50	5	50						
2	22 mm × 15 mm reducers		22		44						
4	8 mm × 6 mm reducers		16		64			2·50			
2	15 mm stop valves	1	45	2	90			0·60			
6	15 mm elbows		08		48			1·80			
2	15 mm × 15 mm × 22 mm tees		32		64			0·60			
2	15 mm screw-on pipe brackets		06		12			0·16			
3 m	15 mm copper tube (B.S. 2871: Part 1)		33		99			0·33			
28·5 m	8 mm microbore copper tube		21	5	99			2·57			
18 m	6 mm microbore copper tube		14	2	52			1·62			
18	8 mm saddle bands and pads		1½		27						
10	6 mm saddle bands and pads		1½		15						
12	8 mm bends (pulled on site)							3·00	13·18	13	18
	Solder, paste and sundries				50	19	14				
	Add to labour:										
	5 % for work on first floor							0·66			
	30 % for work under floors							3·95			
	20 % for work in occupied house							2·64	7·25	7	25
	Carried forward					86	18		51·46	51	46

BILL OF QUANTITIES FOR MICRO BORE CENTRAL HEATING AND HOT WATER SUPPLY IN THREE BEDROOM OCCUPIED HOUSE (Contd.)

Sheet No. 4   *Date:* 25.6.71

	Work on ground floor (radiators, boiler and surface pipework)	Materials						Labour			
		Unit net price						Hours fixing			
		£	p	£	p	£	p	h	h	£	p
	Brought forward					86	18		51·46	51	46
	Steel radiators:										
	Dining room										
	Single panel 2·44 m²	8	15	8	15						
	Lounge										
	Single panel 2·86 m²	9	25	9	25						
	Hall										
	Single panel 1·48 m²	4	90	4	90						
	Kitchen										
	Single panel 0·95 m²	3	20	3	20						
4	radiator air valves		10		40						
10	radiator wall brackets		10	1	00						
2 m	8 mm tube for radiator valve distribution pipes		21		42						
4	twin radiator valves (regulating type)	1	66	6	64	33	96	8·00	8·00	8	00
	Dining room and Lounge radiators add for extra brackets							0·70	0·70		70
1	Natural gas fired small bore boiler unit with pump, electric heating controls and programmer, and gas safety controls. Connect to existing brick flue. Connected to gas and electric supplies	120	00	120	00			5·25			
1	safety valve	4	60	4	60			0·35			
1	combined altitude gauge and thermometer	4	10	4	10	128	70	0·35	5·95	5	95
	Carried forward					248	84		66·11	66	11

BILL OF QUANTITIES FOR MICRO BORE CENTRAL HEATING AND HOT WATER SUPPLY IN THREE BEDROOM OCCUPIED HOUSE (Contd.)

Sheet No. 4 (*Contd.*)   *Date:* 25.6.71

Work on ground floor (radiators, boiler and surface pipework)		Materials						Labour			
		Unit net price						Hours fixing			
		£	p	£	p	£	p	h	h	£	p
	Brought forward					248	84		66·11	66	11
2	28 mm unions		56	1	12			0·70			
2	28 mm tees		32		64			0·70			
1	15 mm elbow		08		08			0·30			
3	28 mm elbows		20		60			1·05			
1	28 mm × 15 mm × 28 mm tee		52		52			0·35			
1·2 m	28 mm copper tube (B.S.2871: Part 1)		76		91			0·18			
3·5 m	15 mm copper tube (B.S.2871: Part 1)		33	1	16			0·39			
5·5 m	22 mm copper tube (B.S.2871: Part 1)		55	3	03			0·72			
2	22 mm screw-on pipe brackets		07		14			0·16			
2	15 mm screw-on pipe bracket		06		06			0·08	4·63	4	63
2	28 mm floor plates		15		30						
2	22 mm floor plates		14		28						
2	22 mm ceiling plates		14		28						
1	15 mm ceiling plate		13		13						
	Solder, paste and sundries				50	9	75				
	Add 10% on boiler time (occupied house)								0·60		60
	Add 20% on pipework and radiator time (occupied house)								2·67	2	67
	Carried forward					258	59		74·01	74	01

BILL OF QUANTITIES FOR MICRO BORE CENTRAL HEATING AND HOT WATER SUPPLY IN THREE BEDROOM OCCUPIED HOUSE (Contd.)

Sheet No. 5   *Date: 25.6.71*

	Work on ground floor (under floor pipework)	Materials						Labour			
		Unit net price						Hours fixing			
		£	p	£	p	£	p	h	h	£	p
	Brought forward					258	59		74·01	74	01
1	Microbore distribution manifold. 22 mm × 8–8 mm branches	4	90	4	90						
2	22 mm × 15 mm reducers		22		44						
4	8 mm × 6 mm reducers		16		64			2·50			
2	15 mm stop valves	1	45	2	90			0·60			
2	28 mm × 15 mm × 22 mm tees		52	1	04			0·60			
6	15 mm elbows		08		48			1·80			
2	28 mm elbows		20		40			0·60			
2	28 mm screw-on pipe brackets		09		18			0·16			
4 m	28 mm copper tube (B.S.2871: Part 1)		76	3	04			0·60			
3 m	15 mm copper tube (B.S. 2871: Part 1)		33		99			0·33			
35 m	8 mm microbore tube		21	7	35			3·15			
9 m	6 mm microbore tube		14	1	26			0·81	11·15	11	15
30	8 mm saddle bands and pads		1½		45						
8	6 mm saddle bands and pads		1½		12						
	Solder, paste and sundries				50	24	69				
	Add the following to labour time										
	30% for under floor work								3·35	3	35
	20% for work in occupied house								2·23	2	23
						283	28		90·74	90	74

SUMMARY OF TENDER
**CONTRACT FOR MICRO BORE CENTRAL HEATING AND CONVERSION
OF EXISTING DIRECT HOT WATER SYSTEM TO INDIRECT OPERATION**

	£	p
(1) Total cost of materials	283	28
(2) Total cost of labour for erection 91 h at £1	91	00
(3) Supervision, testing, cleaning up and operating instruction to householder: 18 h	18	00
(4) Travelling allowance: 14 days at 28p	3	92
(5) National Insurance, Redundancy Fund, Holiday Credits, Sickness and Accident Insurance: 109 h at 11·40p	12	43
(6) Selective Employment Tax: 109 h at 6p	6	54
(7) Cartage of materials: 3 h at £1.40	4	20
(8) Wear and tear of tools: 3 per cent on net erection labour cost	2	73
	422	10
(9) Profit and overheads: Add 15 per cent	63	32
(10) P.C. for electrical work on boiler controls and hot water primary thermostatic control. Includes $2\frac{1}{2}$ per cent for heating contractor	14	00
Total amount of estimate	499	42

Price to tender—£499.

Note: The heat loss from microbore pipe fixed under the floors is not excessive, and in most cases is uninsulated. Where such floors are well ventilated, e.g. with air bricks on more than one side, insulation with sectional flexible foam with split sides jointed with adhesive, is advisable at an extra cost for this three bedroom house of £25, bringing the amended tender to £524.

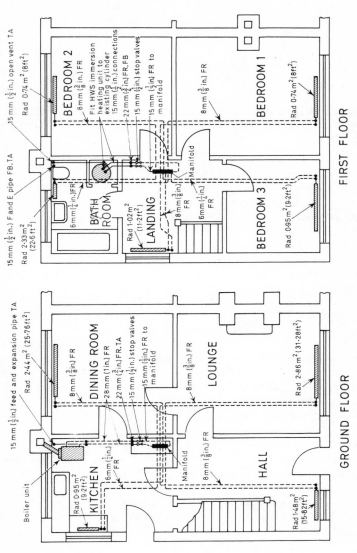

Figure 17. Microbore domestic central heating and hot water supply system in three bedroom semi-detached house

6

NOTES ON COSTING

The use of the priced Bill of Quantities, and the Variation Schedule for the costing of contract variations, have been fully explained in earlier chapters.

Costing Variations

Subject to the Owner's agreement, extras and deductions may be valued at variation schedule rates, or when the work cannot be properly measured, at agreed day work rates. A typical day work rates schedule follows.

Labour (Wages of foreman, fitter, mate)	plus 60 per cent
Statutory labour 'on-costs' (National Health Pensions, S.E.T., Holidays with Pay, Redundancy)	plus 15 per cent
Fares and allowances under National Agreement	plus 15 per cent
Cost of materials and cartage of materials	plus 15 per cent

Note: Percentage additions for day work, and hourly jobbing work rates, are subject to negotiation between the firm and their client. The above day work schedule is given as an example.

On cost of work done for heating contractor by sub-contractors	plus 5 per cent
For sub-contracts containing provision for main contractor of $2\frac{1}{2}$ per cent cash discount, add one-thirty-ninth to the total value of the extra work, calculated at the above rates.	

* Full information on all agreements covering the industry is given in the *Heating and Ventilating Year Book*, H.V.C.A. Coastal Chambers, 172 Buckingham Palace Road, London, S.W.1.

THE COSTING SYSTEM

A simple costing system easily adaptable to all types of work, from repairs and jobbing to large contracts, is essential.

The smaller firms cannot always afford to employ a full-time costing clerk and the costing has to be done by the principal, or by the one clerk-book-keeper, who is responsible for all the clerical and accounting work in the office.

The basic reasons for costing are:

(1) To enable the contractor to know at any stage of the job what his costs are in relation to the work which is completed, and whether or not his profit margin is being maintained. Progress knowledge is vital to the success of the business.

(2) It provides a sound basis for calculating the amount of interim payments, which are generally made each month.

(3) It provides an up-to-date check on the efficiency of the labour engaged on the contract in question, when related to the materials and labour allowed in the Bill of Quantities.

(4) As in (3) it provides a constant check on materials issued to the job. Wastage of materials and labour is the prime reason why contracts sometimes finish on the wrong side. Loss of this kind can only be discovered, and remedied in time, by regular and efficient costing as the work proceeds.

(5) A properly operated costing system allows the final cost to be known immediately on completion of the work, thus avoiding delay in presenting the Final Account.

Labour

Every hour of labour, every metre of pipe, every nut, bolt, and washer, and every fitting, in fact everything no matter how small which is used on the job, should be recorded.

The workman's weekly time sheet is an essential part of the costing system, for without it a true record of the hours worked on the job cannot be kept.

Figure 18 shows a suitable form of time sheet.

The foreman, or chargehand, checks each workman's completed time sheet and countersigns it as being correct, and wages should only be paid for time so completed. All information on the time sheet is checked at the office before wages are paid.

The next step is to enter the hours and wages paid on the weekly cost sheet, of which Figure 19 is a specimen.

Materials

The purchase of materials represents the greatest charge on the capital of a heating contractor. The contractor must therefore ensure that money spent on materials is properly allocated to the job on which they have been used. Failure to book materials to the job means a direct loss, which must of course be avoided.

CONTRACT WORK

In contract work the costing system should record:
     (1) The ordering of the materials.
     (2) Delivery of the materials.
     (3) The use made of the materials.
     (4) Materials put into stock for future use.

All materials should be ordered through the office, the only exception to this rule being the purchase of small items by the foreman on out-of-town contracts. For this purpose the foreman is supplied with a duplicate order book and has the sole responsibility for ordering such items.

Delivery Notes

All materials delivered to the site, whether ordered by head office or by the foreman, must be covered by *Delivery Notes* signed by the foreman or chargehand at the time of delivery. Materials delivered to the firm's stores are signed for by the storekeeper or the cost clerk.

Materials issued by Stores to the Job

An issue note must accompany every consignment of material issued by the firm's central store, and this must be signed by the foreman or chargehand receiving the material on site.

Record of Materials Used

A daily record sheet showing materials issued to the fitters is made up by the foreman, or in the case of a large contract by the site storekeeper.

     Each week the foreman or chargehand posts to the office:

     (1) The time sheets for all labour employed on the site.
     (2) The daily materials record sheets.
     (3) The site duplicate order sheets.
     (4) Delivery notes for all materials brought on to the site.

TIME SHEET FOR WEEK ENDING.............

Day of week	Job or site	Description of work	Started Time	Finished Time	Hours
Wednesday					
Thursday					
Friday					
Saturday					
Sunday					
Monday					
Tuesday					

Workman's signature..............   Allocations and notes.

Foreman's signature...............   Office use only.

Figure 18. Specimen time sheet

The lorry driver's time sheet is, of course, handed in to the office each week and this, together with the four items mentioned above, plus the office order book and the stores issue book, enables the cost clerk to prepare the weekly cost sheet. Totals for each week are carried forward to the next, so that the current sheet gives the up-to-date cost of the job.

As the job proceeds the cost clerk compares the office and foreman's order books and delivery notes with the invoices received. This is to satisfy himself that all the materials charged on the invoices have in fact been ordered and received by his firm.

Unused Materials

Instructions should be given to the foreman to forward to the office a list of all unused materials on the site at the end of the contract. These materials must be entered into stock for use on future contracts.

COSTING JOBBING WORK

Most jobbing work is fed with materials from the contractor's stores. Firms relying mainly on jobbing work maintain stocks of materials such as valves, fittings, tube, and even a selection of radiators, boilers, and hot water cylinders. Spare parts for automatic stokers and oil burners are also held in stock by many heating contractors, who often provide a really good maintenance service for these machines.

In the course of a working week it is possible for one pair of men to complete two or more separate jobs, and care must be taken to see that the correct time, materials, and cartage are correctly allocated to each job.

Materials Transferred from Another Job

If often happens that materials left over from one small job can be used on the next one. In order to keep the office informed the transfer should be recorded by the foreman on his daily materials sheet. If this is done the transfer is known to the office and the materials are charged up to the correct job.

Application for materials from stores should be made by the foreman on an official order addressed to the office, and he should carry a duplicate order book for this purpose. Materials are issued by the storekeeper only on receipt of this order, and the person receiving the materials should sign a *Stores Issue Note*. The storekeeper must be instructed to accept only the signature of a responsible person of no less grade than chargehand.

When the foreman returns materials to the stores at the end of a job he should obtain the storekeeper's signature on a *Stores Returned Note*.

When small additional items such as nuts and bolts, or one or two small pipe fittings, are required for a job, it may be cheaper to

Cost Sheet for week ending ....

Job No. .... Name of job ....

Sheet No. .... Site ....

Nature of Work..

Tender price (if contract) ....

Work started............................. Work completed.............................

Labour						
	Hours	*Rate*				
National Insurance, and Redundancy Fund						
Selective Employment Tax						
Holiday Credit and Sickness and Accident Premiums						
Fares and Allowances						
Abnormal Conditions and/or Danger Money						
Materials						
Cartage						
Sub-contracts						

Figure 19. Specimen weekly cost sheet

purchase these locally than send a man four or five miles to the office. Cash purchases should only be permitted in the circumstances mentioned and the contractor should instruct the workmen to obtain a receipt in every case. The foreman forwards these receipts to the office each week along with the men's time sheets and his daily materials sheets.

Example of Costing for Jobbing Work

Let us suppose that a contractor has received an order to dismantle and remove a 332 litre direct galvanised hot water cylinder, and to fit a new cylinder.

The foreman will visit the job and:

(1) Ascertain the exact nature and extent of the work.
(2) Check the cylinder dimensions and tappings, and whether direct or indirect.
(3) Measure up for any new fittings or mountings required.
(4) Make quite sure that the new cylinder will pass through existing openings and doorways.

The foreman now makes out an order on the firm's stores for the cylinder and all other materials which he considers are needed to complete the work. These materials are issued from the stores and the foreman signs a *Stores Issue Note* and arranges for the firm's lorry to deliver to the site.

The foreman then arranges the programme of work with the building owner, and instructs a fitter and mate when to start work. Tools, and perhaps lifting tackle and blocks, may have to be carted from another job which is nearing completion, in which case the firm's lorry is called upon to do a special journey for this purpose. On the morning work begins the foreman meets the men at the site to give final instructions.

It will be seen that before the actual work starts the contractor has incurred the following expenses:

(1) Foreman's time visiting site, measuring up, preparing list of quantities, drawing materials from stores and arranging transport for both materials and tools to site, arranging labour, and instructing on site.
(2) Storekeeper's issue of materials.
(3) Cost of transporting materials and tools to the site.

Regarding (1), the foreman's time is charged to the job, and entered on the time sheet in the normal way. Item (2) is usually covered in Establishment Charges, and item (3) is charged to the job by the cost clerk at the firm's hourly rate for transport, the time being taken from the driver's time sheet.

If on completion of the work the tools and any unused materials are wanted on a new job, transport for moving will be charged to the new job. If on the other hand the tools and materials must be returned to the contractor's stores, cost of transport is charged to the job from which they are removed.

Figure 20 shows the full costing of the work.

Abnormal conditions money has been allowed for the time taken to dismantle and remove the old cylinder and insulation from the boiler room, which was 6 hours. The allowance is 28p per day or part of a day, and therefore a payment of 28p per man is made and recorded on the time sheet.

The cost sheet records all money paid or due for payment in connection with the finished job, which in this case amounts to £104·30.

As stated in Chapter 5, labour employed on jobbing work is charged at all-in rates, which cover also National Insurance, Redundancy Fund, Holiday Credits, Sickness and Accident Premiums, General Establishment Charges, together with the contractor's due profit on his labour. A charge to cover Selective Employment Tax is also included.

Materials and cartage of materials are charged at net cost plus 20 per cent; all out-working allowances, fares, abnormal conditions money, danger money, and out-of-pocket expenses, plus 20 per cent.

Where the heating contractor employs a specialist firm, e.g. for thermal insulation or electric wiring, he is entitled to an addition of 20 per cent on the amount of the sub-contractor's charge.

For work carried out as an indirect contract containing provision for the usual $2\frac{1}{2}$ per cent cash discount for the main contractor, the heating contractor must add one-thirty-ninth to the total value of the account.

The specimen account (Figure 21), which is based on the above rates, will be presented to the Owner for payment in the amount of £127·03. Referring to Figure 20, we see that the total cost is £104·30, which allows a margin for the heating contractor of £22·73 to cover Establishment Charges and Profit.

INTERIM PAYMENTS FOR CONTRACT WORK

Most contracts carry a clause permitting the contractor to receive, on application, interim payments for materials delivered to site and work done to date. Certificates for interim payment are usually

COST SHEET for week ending _31st July 1971_

Job No. _93_   Name of Job _"Hilltop" Primary School_

                            Site_____Marl Road, Lowtown_

                            Nature of work _Renewal of H.W. Storage Cylinder_

Tender price (if contract) _Day Work_

Work started_26th July 1971_   Work completed_28th July 1971_

LABOUR			HOURS	RATE	£	p
Foreman	J. Shorter					
Fitters	B. Longer		7	65½p	4	59
			25½	55½p	14	15
Mates	E. Cann		25½	44½p	11	35
Selective Employment Tax _Fitter and Mate £1.53 Foreman 42p_					1	95
National Insurance and Redundancy, _Fitter and Mate £2.85_						
Holiday Credits and Accident Premiums _Foreman 42p_					3	27
Fares and Allowances _Fitter and Mate 3 days @ 72p. Foreman_					6	36
Abnormal Work and/or Danger Money _60 miles @ 7p_						56

MATERIALS	£	p				
no 1 – 332 litre, 4mm plate galvanised H.W. Cylinder	25	35				
no 1 – Combined Altitude gauge and thermometer	4	08				
no 2 – 40mm unions @ 87p	1	74				
no 2 – 32mm unions @ 73p	1	46				
no 2 – 25mm unions @ 60p	1	20				
no 2 – 40mm barrel nipples @ 11p		22				
no 2 – 32mm " " @ 9p		18				
no 1 – 25mm " " @ 7p		7				
Jointing Materials	1	25			35	55

CARTAGE						
Own lorry 3 hours @ £1.40	4	20			4	20

SUB—CONTRACTS						
Thermal Insulation	22	32			22	32

| Total net cost | | | | | 104 | 30 |

Figure 20. Specimen cost sheet for jobbing work

Blankshire County Council
County Offices,
Blanktown.

To repairs at Hilltop Primary School, Marl Road, Lowtown.

27 August 1971.

	£	p	£	p
Labour				
Foreman 7 hours at 93p	6	51		
Fitter 25½ hours at 81p	20	66		
Mate 25½ hours at 67p	17	08	44	25
Fares and Allowances				
Foreman 3 car journeys. 60 miles at 7p	4	20		
Fitter and Mate 3 days at 72p	2	16		
Add 20%	1	27	7	63
Abnormal Work Allowance				
Fitter and Mate 6 hours		56		
Add 20%		11		67
Materials				
New cylinder, mountings, pipes and fittings	35	55		
Add 20%	7	11	42	66
Cartage				
Lorry 3 hours at £1.40	4	20		
Add 20%		84	5	04
Sub-contract				
Insulation of primary pipes and cylinder	22	32		
Add 20%	4	46	26	78
			127	03

Figure 21. Specimen account for jobbing work

issued by the architect or by the consulting engineer at monthly intervals.

The heating contractor, if engaged upon a direct contract, will make monthly application to the owner or his architect. In the case of a sub-contractor the application is addressed to the general contractor, who is responsible for payments to his sub-contractors, subject to certification by the architect or the consulting engineer.

Each month the heating contractor submits a statement giving the value of work completed and of materials on site. Full information for valuing the work can, of course, be obtained from the cost sheets.

A formal application for interim payment should give the value of the accepted tender, the total amount of any payments already received, and the value of the present application.

The architect or consulting engineer examines the application and if satisfied with the contractor's valuation will issue a certificate for payment.

RETENTION MONEY

Ninety per cent of the value of the work and materials is usually certified for payment within fourteen days. The remaining 10 per cent, termed the *retention money*, is payable as follows:

Five per cent is due upon final completion of the work, and the remaining 5 per cent at the end of the Guarantee or Defects Liability period; this is usually twelve months after final completion and acceptance of the installation by the owner.

To illustrate the application of the above, consider a heating contract valued at £10 000.

Available for heating contractor in monthly interim payments during the run of the contract = 90 per cent of £10 000		£9 000
5 per cent payable on completion of work		500
5 per cent payable at end of Guarantee Period		500
	Total	£10 000

The date of handing over of the installation to the owner marks the beginning of the Guarantee Period, during which the heating contractor must make good any defective materials and/or workmanship which may have become evident. When the work is complete the heating contractor should write to the owner or to his agent, informing him of this and indicating that the Guarantee Period will start on that date and will terminate on a date 12 months hence. A form of letter dealing with these matters would be:

9.8.71

Dear Sirs,

Heating of New Laboratories.

The whole of my work on the above-named contract is now complete and tests have been carried out to the satisfaction of your Engineer.

I shall be pleased to hear that you agree that the Defects Liability Period shall begin from the date of this letter and terminate on 9 August 1972.

Will you also please arrange my immediate release from the bond in connection with this contract.

Yours faithfully,

RELEASE FROM THE BOND

Under the terms of the contract the heating contractor may be required to provide, at his own expense, a bond as security for due performance of an amount equal to 5 per cent of the contract sum. The bond will be arranged with a bank or an insurance company or in the form of cash.

On a large contract maintenance of the bond is quite an expensive matter for the contractor and therefore he should make a formal request for release as early as possible after completion and testing of the installation. This request can be included in the letter referred to above.

FINAL PAYMENT

As explained earlier in this chapter, upon completion of the work and its acceptance by the owner, the contractor is entitled to receive payment up to the value of 95 per cent of the final cost. The remaining 5 per cent is retained by the owner until the end of the Defects Liability Period, when the contractor should apply for final settlement.

The reader is advised to study the R.I.B.A. Form Of Main Contract, and where the heating sub-contractor is nominated under this form of contract, the NFBTE/FASS/CASEC Standard Form Of Sub-Contract should be used. Full information regarding these documents is contained in the current edition of the Heating and Ventilating Yearbook. (See Appendix A.)

7

BUILDER'S WORK

When quoting direct to the owner for heating and ventilating work, items of builder's work may, at the owner's request, be included as a separate item. This will normally be a quotation from a builder to the heating contractor, based upon a schedule of items prepared by the latter. Should the builder's work not be included in the heating contractor's tender he will make this clear to the owner and give a schedule of the building work which has been excluded.

In the case of a new building the invitation to tender is usually received from the architect. When tendering the heating contractor will enumerate the items excluded from his tender and give a full schedule of builder's work necessary for the successful completion of the heating and ventilating installations. The architect will forward the builder's work schedule and the tender price to the quantity surveyor for inclusion in the main Bill of Quantities.

If a consulting engineer is employed to design and supervise the engineering services, he will prepare the schedule of builder's work, and also supply dimensions, with drawings when required, of all spaces and structures such as boiler rooms, pump rooms, fan chambers, fuel storage, boiler bases, fan bases, stoker and oil burner bases, chimneys and brick flues, subways, trenches, sumps, etc.

Much of this information, especially that concerning plant rooms, fuel storage, and trenches, may be required by the architect during the early stages of planning.

The Schedule of Builder's Work

The schedule should state clearly and precisely every item of builder's work required. Minimum dimensions must be stated, and in the case of boiler and plant rooms and unusually-shaped machine bases, dimensioned drawings must be provided for the guidance of the architect and the quantity surveyor when planning and pricing.

A typical schedule is given on pages 223, 224 and 225.

SCHEDULE OF BUILDER'S WORK IN CONNECTION WITH PROPOSED HEATING, HOT WATER SUPPLY, AND VENTILATION

at

Name of Building ....

Address ....

....

(1) Provide Boiler Room to dimensions given on drawings Nos. ......

(2) Provide Fuel Store to dimensions and design given on drawings Nos. ......

(3) Provide brick chimney to the following dimensions: *Height* ...... m measured from Boiler Room floor level. Internal dimensions to be ...... mm × ...... mm in the clear. The chimney to be lined to full height with molar bricks bonded with the outside brickwork.

(4) Provide horizontal flue in brickwork, lined as above, to the following requirements in clear dimensions: ...... mm × ...... mm.

(5) Provide and build in No. ...... cast iron flue cleaning doors and frames. *Size* ...... mm × ...... mm.

(6) Build into horizontal flue No. ...... steel sleeves, to be provided by heating contractor.
No. ...... for heating boilers, at ...... mm inside diameter.
No. ...... for hot water supply boiler at ...... mm inside diameter.

(7) Build in No. 1 ...... mm × ...... mm draught stabiliser, to be provided by heating contractor.

(8) Provide No. ...... boiler bases to dimensions on drawing No. ...... Construct on 100 mm insulation concrete foundation, with bullnosed blue brick on edge, and 225 mm × 225 mm bullnosed corner blocks.

(9) Form No. 2 smooth concrete bases, ... mm × ...... mm × ......mm high, for hot water cylinders.

(10) Form No. 4 smooth concrete pump bases ...... mm × ...... mm × ...... mm high. Prepare holes for holding-down bolts to template provided by heating contractor.

(11) Form sump 0·6 m × 0·6 m × 0·8 m deep in Boiler Room floor

for drainage pump, and supply and fix channel or angle iron pump supports across sump, to Engineer's instructions on site.

(12) Provide fan chamber of dimensions and in position shown on Plan No. ......

(13) Cut away for and pin into walls No. ...... rag bolts. (*Give sizes.*)

(14) Form smooth concrete base for centrifugal fan and motor, to details on drawing No. ......

(15) Cut No. ...... holes through 112·5 mm brick walls and build in pipe sleeves.

(16) Cut No. ...... holes through 225 mm brick walls and build in pipe sleeves.

(17) Cut No. ...... holes through ...... floors and build in pipe sleeves. (*Describe thickness and type of floor.*)

(18) Cut away for and pin in No. ...... schoolboard pattern pipe brackets in rooms.

(19) Cut away for and pin in No. ...... purpose-made pipe hangers and supports in Boiler Room.

(20) Cut away for and pin in No. ...... purpose-made pipe brackets and supports in subways, trenches, and ducts. (*Describe types and sizes.*)

(21) Form pipe trenches and ducts with removable covers, to lengths and cross-sectional area in clear dimensions, as shown on drawings Nos. ......

(22) Provide subways, with access manholes, ventilation, and drainage, to lengths and cross-sectional area in clear dimensions, indicated on drawing No. ......

(23) Cut away for and pin in No. ...... radiator wall brackets, and No. ...... radiator top stays.

(24) Fix No. ...... radiator wall shields.

(25) Provide and lay concrete or timber supports for cold water storage tanks, and for heating feed and expansion tank.

(26) Box in the above tanks with 13 mm timber, complete with wooden covers. Leave 50 mm space between tanks and boxing, and fill in with granular cork or glass silk insulating materials.

(27) Provide holes through roof, or through Tank Room walls, for overflow pipes from tanks, and fit lead slates for pipes.

(28) Lay on cold water supply to tanks in lead and connect to tank ball valves. Provide and fit separate stopcock to each tank cold water supply pipe connection.

(29) Paint approximately ...... m² of radiator surface.

(30) Paint approximately ...... m² of pipe surface.

(31) Build in ventilating registers and fresh air inlet gratings as follows: (*Indicate number, location, and sizes.*)

(32) Allow Engineer the use of ladders, trestles, platforms, and scaffolding during progress of work.

(33) Attend on Engineer during the progress of the work.

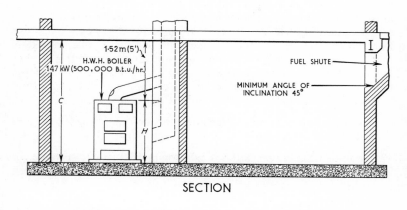

SECTION

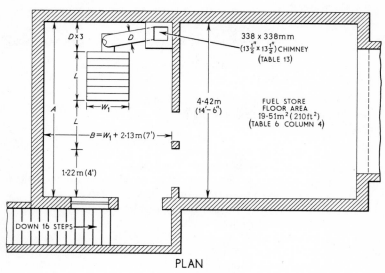

PLAN

Figure 22. Boiler house dimensions (single boiler)

Sizing the Boiler Room

More often than not the shape of the boiler room will be dictated by the structural and planning requirements of the architect.

On the other hand the engineer's concern is to arrange his plant so that it can be easily and efficiently operated and maintained. In order to secure these conditions adequate floor area and height are necessary, and the engineer must be prepared at a very early stage in the design to inform the architect of the minimum dimensions of the proposed boiler room, fuel store, and chimney. With this information before him the architect can arrange to accommodate the engineering plant within his scheme. Some compromise has invariably to be agreed between architect and engineer, and the final result may allow more or less space than the engineer would like for what he considers to be an ideal plant arrangement.

Minimum boiler room dimensions can be determined by use of the following formulae:

For housing a single boiler as shown in Figure 22; suitable for hand-fired coke, bunker to boiler automatic underfeed coal stoker, or oil burner installations:

$A=[(L\times2)+(D\times3)]+1\cdot22$ m.
$B=W_1+2\cdot13$ m.
$C=H+1\cdot52$ m.

Note: Add $1\cdot0$ m to A when hopper type automatic stokers are used.

The space required for housing more than one boiler, as shown in Figure 23, is given by the formula:

$A=[(L\times2)+(D\times3)]+1\cdot22$ m.
$B=(W_1+W_2+1\cdot52$ m$)+(N\times0\cdot61$ m$)$
$C=H+1\cdot52$ m.

Where A=length of boiler room;
     B=width of boiler room, for boilers only;
     C=height of boiler room, measured from finished floor level to underside of beams, or joists;
     D=diameter of largest boiler smoke outlet;
     H=height of highest boiler;
     L=length of largest boiler;
     N=number of boilers;
     W_1=width of one boiler (W_2, W_3, W_n, for widths of successive boilers).

Note: Add $1\cdot0$ m to A when hopper type automatic stokers are used.

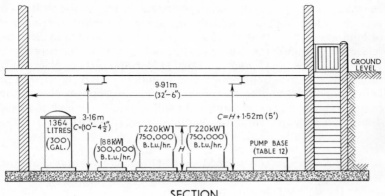

SECTION

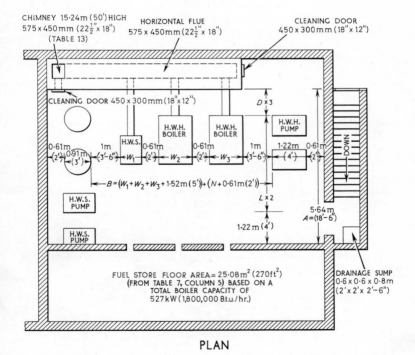

PLAN

Figure 23. Boiler room dimensions (boilers, cylinders and pumps)

Provision for hot water cylinders and pumps, where these are to be sited in the boiler room, will necessitate adding to the width. To accommodate this plant, add the net space required to each piece of equipment, plus 0·61 m, to dimension *B*. This is made clear by reference to Figure 23, where on one side of the boiler room two 1364 litres H.W.S. cylinders and two H.W.S. pumps are placed; for these we add the diameter of one cylinder = 1·0 m, plus 0·61 m between cylinder and boiler room wall. On the opposite side of the boiler room two heating pumps are to be installed, which increase dimension *B* by the length of the pump base, plus 0·61 m.

Boiler Rooms for Gas-fired Systems. Dimensions *B* and *C* are the same for gas boilers as for solid fuel and oil installations.

The formula for dimension *A* is:

$$A = (L \times 2) + (D \times 3).$$

No fuel storage is needed, but provision must be made for housing the gas meter in a separate brick or concrete chamber, to the following internal dimensions:

Meter capacity m^3/h	Height		Length		Width	
	m	*ft*	*m*	*ft*	*m*	*ft*
35 to 55	1·83	6·0	1·90	6·25	1·50	5·0
85 to 130	2·00	6·5	2·20	7·25	1·70	5·5
170	2·44	8·0	2·70	8·75	1·83	6·0
255	2·44	8·0	3·00	9·75	2·00	6·5

Nitrogen-operated Hot Water Heating Pressurising Plant. The space required for this equipment is:

Rating of plant		Length		Width		Height	
kW	*Btu/h*	*m*	*ft*	*m*	*ft*	*m*	*ft*
Up to 879	3 000 000	3·35	11	3·00	10	3·00	10·0
Up to 2 931	10 000 000	5·50	18	3·66	12	3·80	12·5
Up to 5 862	20 000 000	7·62	25	3·66	12	4·60	15·0

Fuel Storage

The amount of fuel to be stored will depend upon the type of build-ing. Hospitals and nursing homes must of course be given special consideration, in order to ensure that fuel is always available, and stock for four to six weeks' consumption at full load is the usual aim. To provide this amount of storage some portion of the stock has often to be accommodated outside the building.

The needs of most buildings will be satisfied if two weeks' con-sumption at full load is provided, i.e., sufficient fuel to sustain the full heat requirements during what may be the coldest fourteen days of the year. For the purpose of the estimate it must be assumed that during this period the outside temperature will fall below the basic design temperature, and consequently the whole boiler plant may be working at its full rated capacity in order to cope with the overload.

For the purpose of fuel store sizing buildings can be divided into two classes:

Class 1. Buildings which require full heat for 24 hours each day, such as hostels, police stations, fire stations, nursing and maternity homes, residential buildings, etc.

Class 2. Those buildings which require full heat during the day-time only, and which are on part load during the night time. These include day schools, offices, certain factories, showrooms, libraries, and business premises generally.

Fuel storage for Class 1 buildings is based on 14 days at full load, and Class 2 on 14 days at 75 per cent full load. For this purpose the term *full load* is taken to mean the rated output of all boilers installed.

Tables 6 and 7 give the tonnage and floor area required for storing coke and small coal. Table 8 gives the quantities of fuel oil to be stored, and Table 9 the dimensions of fuel oil storage tanks. These tables contain all the information required for fuel store sizing.

Figure 24 illustrates the setting of the fuel oil storage tank in the tank chamber, minimum clearances between tank and structure, and tank supports. Additional oil fuel tank sizes can be obtained from manufacturers' catalogues.

Cold Water Storage

Minimum tank room dimensions can be taken from Figure 25. Water tanks which vary considerably in construction and size for a given capacity should be sized from the makers' catalogues. Some water tank sizes are given in Table 10.

Table 6. FUEL STORAGE FOR BUILDINGS IN USE 24 HOURS PER DAY

Total boiler capacity		Capacity of fuel store Tonnes (metric)		Floor area m²	
kW	1000 Btu/h	Coke	Coal	Coke	Coal
29	100	2·5	2·0	3·90	1·86
59	200	5·0	4·0	7·80	3·72
88	300	7·5	6·0	11·71	5·57
117	400	10·0	8·0	15·61	7·43
147	500	12·5	10·0	19·51	9·29
176	600	15·0	12·0	23·41	11·15
205	700	17·5	14·0	27·31	13·01
234	800	20·0	16·0	31·21	14·86
264	900	22·5	18·0	35·12	16·72
293	1000	25·0	20·0	39·02	18·58

Basis of Table

Coke: C.V. 26 000 kJ/kg.   55 per cent efficiency.   Density 433 kg/m³
Coal: C.V. 28 000 kJ/kg.   65 per cent efficiency.   Density 721 kg/m³
14 days' consumption at full load.
Stacking height of fuel 1·52 m (5 ft).

Table 7. FUEL STORAGE FOR BUILDINGS IN USE DURING DAYTIME ONLY

Total boiler capacity		Capacity of fuel store Tonnes (metric)		Floor area m²	
kW	1000 Btu/h	Coke	Coal	Coke	Coal
29	100	2·0	1·5	3·0	1·40
59	200	4·0	3·0	6·0	2·80
88	300	6·0	4·5	9·0	4·20
117	400	8·0	6·0	12·0	5·60
147	500	10·0	7·5	15·0	7·00
176	600	12·0	9·0	18·0	8·40
205	700	14·0	10·5	21·0	9·80
234	800	16·0	12·0	24·0	11·20
264	900	18·0	13·5	27·0	12·60
293	1000	20·0	15·0	30·0	14·00

Basis of Table

Coke: C.V. 26 000 kJ/kg.   55 per cent efficiency.   Density 433 kg/m³
Coal: C.V. 28 000 kJ/kg.   65 per cent efficiency.   Density 721 kg/m³
14 days' consumption at 75 per cent full load.
Stacking height of fuel 1·52 m (5 ft).

Table 8. OIL FUEL STORAGE
A. Buildings in use 24 hours per day.   **B.** Buildings in use during daytime only.

Total boiler capacity		Capacity of fuel tanks (10% has to be allowed for sludge, etc.)			
		litres		gallons	
kW	1000 Btu/h	A	B	A	B
29	100	1 370	1 020	300	225
59	200	2 730	2 050	600	450
88	300	4 100	3 070	900	675
117	400	5 460	4 100	1 200	900
147	500	6 830	5 120	1 500	1 125
176	600	8 190	6 140	1 800	1 350
205	700	9 560	7 170	2 100	1 575
234	800	10 920	8 190	2 400	1 800
264	900	12 290	9 220	2 700	2 025
293	1 000	13 650	10 240	3 000	2 250

Basis of Table

200 sec oil   C.V. 40 800 kJ/litre, 70 per cent efficiency.
A. 14 days' consumption at full load.
B. 14 days' consumption at 75 per cent full load.

Table 9. DIMENSIONS OF FUEL OIL TANKS

Approx. capacity		Rectangular tank			Cylindrical tanks	
litres	gallons	Length m	Width m	Depth m	Length m	Internal diameter m
455	100	0·762	0·762	0·762		
910	200	1·00	1·00	1·00		
1 370	300	1·12	1·12	1·12		
1 820	400	1·22	1·22	1·22		
2 280	500	1·52	1·22	1·22	1·75	1·37
2 840	625	1·91	1·22	1·22	2·01	1·37
3 410	750	2·29	1·22	1·22	2·44	1·37
4 550	1 000	2·44	1·52	1·22	3·28	1·37
5 690	1 250	2·44	1·52	1·52	3·35	1·52
6 830	1 500	2·44	1·83	1·52	2·82	1·83
7 960	1 750	2·44	1·83	1·83	3·35	1·83
9 100	2 000	2·44	2·44	1·52	3·28	1·98
11 380	2 500	3·05	2·44	1·52	4·65	1·83
13 650	3 000	3·05	2·44	1·83	4·80	1·98
18 200	4 000	3·05	2·44	2·44	4·72	2·29
22 750	5 000	3·96	3·05	2·44	5·94	2·29

Note: To calculate approximate weight of tank contents allow 1 litre = 0·93 kg.

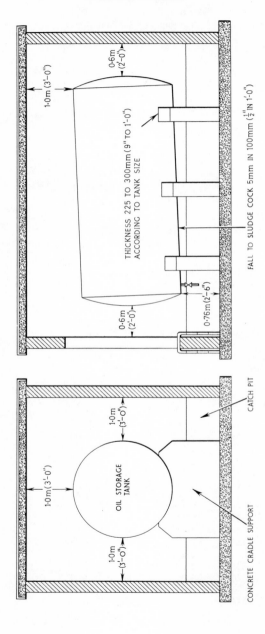

Figure 24. Typical details of fuel oil storage tank installation

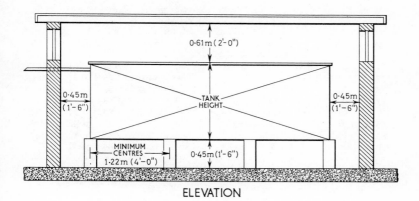

ELEVATION

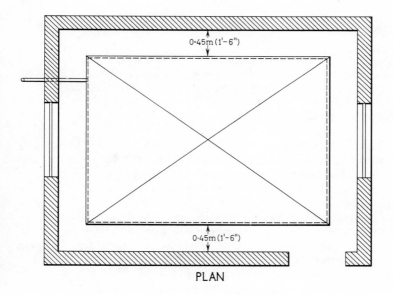

PLAN

Figure 25. Minimum dimensions for cold water storage tank rooms

Table 10. OUTSIDE DIMENSIONS OF OPEN TOP WATER TANKS

Nominal capacity		Metres			Nominal capacity		Metres		
litres	gallons	L	W	H	litres	gallons	L	W	H
182	40	0·689	0·508	0·508	1 140	250	1·52	0·914	0·813
227	50	0·738	0·559	0·559	1 360	300	1·83	0·914	0·813
273	60	0·762	0·584	0·610	1 590	350	1·52	1·14	0·914
318	70	0·914	0·584	0·610	1 820	400	1·83	1·08	0·914
364	80	0·914	0·660	0·610	2 280	500	1·83	1·22	1·02
455	100	1·22	0·610	0·610	2 730	600	1·83	1·22	1·22
682	150	1·09	0·864	0·737	3 640	800	2·44	1·22	1·22
910	200	1·17	0·940	0·889	4 550	1 000	2·44	1·52	1·22

Note: To convert metres to millimetres multiply by 1 000.

Boiler Bases

A typical brick base for a sectional boiler is shown in Figure 26. The dimensions A and B will depend upon the make of boiler specified, and can be obtained from the boiler makers' catalogue.

Hot Water Cylinder Bases

Concrete or brick bases for cylinders fixed vertically are usually formed square, each side being equal to the cylinder diameter, plus 300 mm. This allows a minimum 150 mm margin for support.

For example, a cylinder of 900 mm diameter requires a 1 200 mm square base.

The height of the base will be arranged to facilitate the connection of the primary return pipe from the boiler or, in the case of a steam calorifier, the piping of the condensate to the steam trap. The base height does not usually exceed 250 mm.

Horizontal cylinders need cast iron or steel cradles, mounted on concrete or brick supports, the number required depending upon the cylinder length. An approximate rule is to allow for supports at 4 ft centres.

The above rule also applies to cylinders supported on cantilever wall brackets, or combined wall and floor supports.

For dimensions of hot water cylinders see Table 11, p. 236.

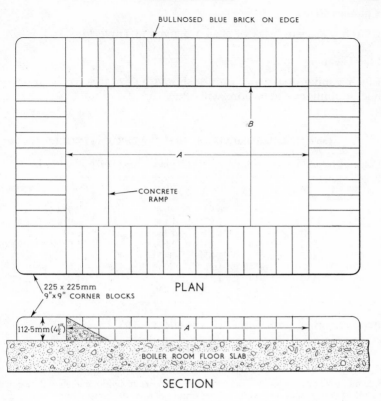

BULLNOSED BLUE BRICK ON EDGE

B

A

CONCRETE
RAMP

225 x 225mm
9"x9" CORNER BLOCKS

PLAN

112·5mm(4½")

A

BOILER ROOM FLOOR SLAB

SECTION

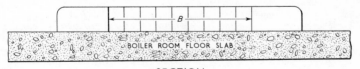

B

BOILER ROOM FLOOR SLAB

SECTION

Figure 26. Typical brick base for sectional boiler

Pump Bases

Brick or concrete bases for pumps will vary in size according to the make and type of pump. The final dimensions will be determined by the machine base plate size and by the pipe connections. When preparing a schedule of builder's work the dimensions given in Table 12 may be used.

Chimney Sizes

Chimney sizes in brick or steel are given in Table 13.

Flue Cleaning Doors

Flue cleaning doors and frames should be as large as the internal flue or chimney dimensions will allow.

Table 11. DIMENSIONS OF HOT WATER CYLINDERS

Approx. capacity		Metres		Approx. capacity		Metres	
litres	gallons	diameter	length	litres	gallons	diameter	length
159	35	0·457	0·99	910	200	0·914	1·58
205	45	0·457	1·22	1 140	250	0·914	1·88
273	60	0·508	1·32	1 360	300	0·914	2·18
364	80	0·610	1·22	1 590	350	1·07	1·93
455	100	0·610	1·63	1 820	400	1·07	2·03
568	125	0·914	1·38	2 050	450	1·07	2·39
682	150	0·914	1·52	2 280	500	1·22	2·13
796	175	0·914	1·83	2 730	600	1·22	2·56

Note: To convert metres to millimetres multiply by 1 000.

Table 12. APPROXIMATE DIMENSIONS OF HEATING PUMP BASES

The final size of the base will depend upon the make of pump, and should be equal to the overall length and width of the frame, plus a 150 mm margin on all sides.

Size of suction and delivery branches (bore)		Horizontally mounted direct-coupled pumps, and belt-driven overhung pump with separate motor				Vertically mounted belt driven pump			
		Length		Width		Length		Width	
mm	in	m	ft	m	ft	m	ft	m	ft
25	1	0·914	3·0	0·610	2·0				
40	1½	1·07	3·5	0·610	2·0				
50	2	1·07	3·5	0·610	2·0				
80	3	1·22	4·0	0·762	2·5	0·610	2·0	0·457	1·5
100	4	1·37	4·5	0·762	2·5	0·762	2·5	0·533	1·75
150	6	1·37	4·5	0·762	2·5	0·914	3·0	0·610	2·0

Note: The height of the base will be dictated by site requirements. For the purpose of the builder's work schedule a height of 450 mm can be given for all sizes.

Table 13. RECOMMENDED MAXIMUM CAPACITIES OF CHIMNEYS FOR
H.W.H. AND H.W.S. INSTALLATIONS
kW

Chimney dimensions (millimetres)			Metal chimney, internal diameter, millimetres									
			250/ 300	350	400	450	500	550	600	650	700	810
			Brick chimney to brick sizes, millimetres Internal dimensions (approx.)									
			225 × 225	338 × 225	338 × 338	450 × 338	450 × 450	562 × 450	562 × 562	675 × 562	675 × 675	788 × 675
Height of chimney in metres from centre of boiler smoke connection to top of chimney stack	6	56	79	130								
	9	73	107	161	215	288						
	12	82	125	188	252	336	417	522	624			
	15		139	211	278	374	463	586	703	844		
	18			227	308	406	504	633	759	910	1 080	
	24				355	472	586	733	882	1 060	1 230	
	30					586	654	821	1 150	1 180	1 380	

Fan Bases

The length and width of centrifugal fan bases vary considerably according to the make, type and size of machine.

Table 14 includes approximate dimensions of bases for the more widely used sizes of fan and will enable preliminary sizes of fan chambers to be determined. The height of ductwork entering and leaving the fan chamber will of course decide the chamber height.

Subways for Heating and Water Services

Main subways for the larger schemes should have a minimum height and width of 2 m.

Crawlways for Heating and Water Services

The minimum dimensions of crawlways carrying one pair of heating mains up to 150 mm bore, and hot water and cold water distribution mains, should be 1·37 m high and 1·22 m wide, in clear.

Crawlways to accommodate the branches of these services should not be less than 1·10 m high and 1·10 m wide, in clear.

Figure 27 shows an arrangement of main services in a crawlway.

Table 14. APPROXIMATE DIMENSIONS FOR CENTRIFUGAL FAN BASES IN CONCRETE OR BRICK

Approx. diameter of fan inlet		Direct coupled fan and motor with overhung runner or belt driven fan without motor		Direct coupled fan and motor with intermediate bearing	
		Length	Width	Length	Width
mm	in	metres	metres	metres	metres
300	12	1·00	0·80	1·40	0·80
375	15	1·10	0·80	1·50	0·80
450	18	1·20	1·00	1·70	1·00
600	24	1·40	1·00	2·00	1·00
750	30	1·50	1·10	2·10	1·10
900	36	1·80	1·10	2·70	1·10
1 200	48	2·10	1·80	3·00	1·80
1 350	54	2·40	2·00	3·50	2·00
1 500	60	3·00	2·40	4·50	2·40

Note: The height of the base will be dictated by site requirements. For the purpose of the builder's work schedule a height of 300 mm can be given for all sizes. To convert metres to millimetres, multiply by 1 000.

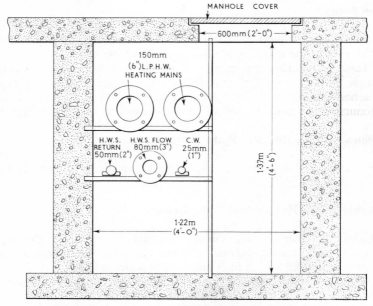

Figure 27. Typical arrangement of main services in crawlway

Manholes for inspection and maintenance, and adequate natural ventilation are necessary.

Pipe Trenches and Ducts

Tables 15 and 18 inclusive give cross-sectional dimensions in clear for pipe trenches and vertical ducts.

Table 15. PIPE TRENCHES FITTED WITH MOVABLE COVERS. MINIMUM CROSS-SECTIONAL DIMENSIONS MM IN CLEAR

For straight pipes with **screwed** joints.
No allowance made for valves or branches.

Bore of pipe		Two pipes				Single pipe			
		Insulated		Uninsulated		Insulated		Uninsulated	
mm	in	Width	Height	Width	Height	Width	Height	Width	Height
15	$\frac{1}{2}$	260	205	230	150	205	205	180	150
20	$\frac{3}{4}$	260	205	230	150	205	205	180	150
25	1	300	205	230	180	230	205	205	180
32	$1\frac{1}{4}$	300	205	230	180	230	205	205	180
40	$1\frac{1}{2}$	360	230	260	205	260	230	230	205
50	2	410	230	300	205	260	230	230	205
65	$2\frac{1}{2}$	460	260	330	230	285	260	230	230
80	3	485	280	360	230	285	285	230	230
100	4	530	300	385	260	300	300	260	260

Note: For a given length of trench, increase the depth by the amount of the pipe gradient.

Table 16. PIPE TRENCHES FITTED WITH MOVABLE COVERS. MINIMUM CROSS-SECTIONAL DIMENSIONS MM IN CLEAR

For straight pipes with **flanged** joints.
No allowance made for valves or branches. (Insulated or uninsulated)

Bore of pipe		Two pipes		Single pipe	
mm	in	Width	Height	Width	Height
65	$2\frac{1}{2}$	560	330	410	330
80	3	585	360	410	360
100	4	640	385	435	385
125	5	710	460	460	460
150	6	965	510	510	510

Note: For a given length of trench, increase the depth by the amount of the pipe gradient.

Table 17. VERTICAL PIPE DUCTS. MINIMUM CROSS-SECTIONAL DIMENSIONS MM IN CLEAR

For straight pipes with **screwed** joints.
No allowance made for valves or branches.

Bore of pipe		Two pipes				Single pipe			
		Insulated		Uninsulated		Insulated		Uninsulated	
mm	in	Width	Depth	Width	Depth	Width	Depth	Width	Depth
15	$\frac{1}{2}$	230	150	150	100	205	150	125	100
20	$\frac{3}{4}$	230	150	150	100	205	150	125	100
25	1	230	150	150	100	205	150	125	100
32	$1\frac{1}{4}$	260	150	180	100	205	150	125	100
40	$1\frac{1}{2}$	300	180	205	125	230	180	150	125
50	2	360	180	230	125	260	180	180	125
65	$2\frac{1}{2}$	385	205	260	125	260	205	180	125
80	3	410	230	285	150	260	230	180	150
100	4	460	260	300	180	285	260	205	180

Table 18. VERTICAL PIPE DUCTS. MINIMUM CROSS-SECTIONAL
DIMENSIONS MM IN CLEAR

For straight pipes with **flanged** joints.
No allowance made for valves or branches. (Insulated or uninsulated)

Bore of pipe		Two pipes		Single pipe	
mm	in	Width	Depth	Width	Depth
65	2½	435	260	285	260
80	3	460	285	285	285
100	4	530	300	300	300
125	5	600	360	360	360
150	6	650	385	385	385

These are for pipes run in open trenches, and ducts fitted with full length removable covers. Figure 28 shows the arrangement of pipes in the trench.

If valves are fitted or branches taken from the pipes, the depth of the trench must be increased to accommodate the largest branch, or the height of the largest valve on the run of pipe.

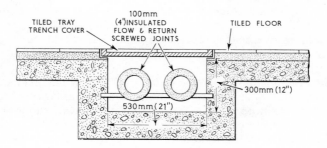

Figure 28. Typical arrangement of pipes in trench

8

ESTIMATING ANNUAL RUNNING COSTS

A statement giving the probable annual running costs of the proposed installation is often asked for by the building owner or his architect.

The owner may also require information about the capital and running costs of other types of installation before coming to a final decision.

The various items of expenditure which make up the annual cost of operating a heating installation can be classified under two main headings:

(1) *Running costs*, which include fuel, cost of maintenance and repairs, electricity for motors, and boiler room lighting.

(2) *Standing charges*, which include labour for attendance, repayment of capital and interest on loan (or an annual sum for depreciation) for both plant and boiler house building, insurance of plant and building, overhead charges to cover cost of ordering fuel, spares, cost of supervision, payment of wages, and other expenditure in connection with accounting.

When making a comparison of annual costs, incidental expenditure will often affect the result. For example, damage to internal decorations is more pronounced where free-standing radiators and convectors are employed than is the case with concealed ceiling or floor heating. On this point no definite rule can be stated, but observations have shown that for certain types of large public building the period between complete internal decorations can be extended from five years in the case of radiators to seven years where floor heating is in use.

In a large building which costs £7 000 to decorate internally the saving is considerable and must be taken into account when comparing the costs of different methods of space heating.

The amount of fuel consumed depends upon several factors, the variable nature of which renders a really close prediction extremely difficult, and an accuracy of ±10 per cent can be considered satisfactory where the forecast is based upon calculations. A much closer estimate is possible when the new building can be compared with buildings of the same type, usage, construction and heat load, with heating plant which has been operating for a few years and for which average annual fuel deliveries are known. Unfortunately, information of this precise nature is costly to obtain, and is not freely available. The heating engineer will therefore make the best use of published data.

ESTIMATING FUEL CONSUMPTION

If all the heat transmitted from the furnace to the water in a heating boiler could be used to perform useful work in warming the building, a close estimate of the maximum hourly fuel consumption would be given by the expression:

$$\frac{\text{Maximum fuel}}{\text{used per hour}} = \frac{\text{Calculated building heat loss}}{\text{Calorific value of fuel} \times \text{boiler efficiency}}$$

In practice, however, this does not happen, because of heat losses which occur from the heating mains and from other causes such as poor design (involving overheating, badly planned distribution, inefficient thermostatic control), bad building design and construction, and last but not least, habitual and unnecessary overheating and excessive ventilation (windows left wide open, etc.) by the occupants of the building. Mains losses can be limited by careful design and effective insulation, but many of the factors which contribute to utilisation inefficiency are beyond the designer's control, and furthermore their effect upon the overall efficiency is unpredictable.

When one considers also the variable quality of boiler attendance and maintenance, the difficulty of accurately predicting the seasonal fuel consumption will be realised.

It is a fact that the great majority of heating and hot water supply installations suffer from the effects of inefficient or unreliable attendance.

The best results are obtained from installations which are in the care of properly trained full-time boiler attendants, under the supervision of plant engineers.

The care given to an installation will be reflected in its performance, and although results are not precisely measurable, it can be safely assumed that the average overall efficiency of a heating installation which is attended by a skilled boiler attendant will be 5 per cent than a similar plant in the care of the average building caretaker.

Overall Efficiency

The utilisation losses determine the *overall efficiency* of an installation and therefore a more accurate assessment of the maximum hourly fuel consumption is:

$$\text{Maximum fuel consumption per hour} = \frac{\text{Calculated building heat loss}}{\text{Calorific value of fuel} \times \text{overall efficiency}}$$

With a hot water supply installation, the utilisation losses include some of those already mentioned, in addition to waste of hot water; thus:

$$\text{Fuel consumption per hour} = \frac{\left(\begin{array}{l}\text{Estimated daily heat for hot water} + \text{Average daily heat loss from hot water pipes in rooms}\end{array}\right)}{\text{Calorific value} \times \text{Overall efficiency} \times \text{Hours per day plant is in use}}$$

Table 19 gives overall efficiencies recommended for estimating the maximum hourly fuel consumed in well-designed heating systems.

Table 19. AVERAGE OVERALL EFFICIENCIES OF SPACE HEATING AND HOT WATER SUPPLY INSTALLATIONS

Heating medium or fuel used	Overall efficiency (per cent)	
	Installation in charge of engineer	Installation in charge of caretaker
Coke (hand-fired)	55	50
Coal (auto-stoker)	65	60
Oil	70	65
Gas	75	70
Electric hot water circulator	95 (which includes heat loss from circulator surfaces)	
Electric radiators, convectors, reflectors	100	

Note: The average overall efficiency of installations fired by modern automatic magazine coke boilers can be as high as 75 per cent.

Calorific Values of Fuels

The calorific value of both gas and furnace coke, and small coal varies between 23 400 kJ/kg (10 000 Btu/lb) and 30 200 kJ/kg (13 000 Btu/lb).

Fuel oils suitable for heating boilers have an average calorific value of 39 400 kJ/litre, (170 000 Btu/gal).

The calorific value of town gas varies between 14 900 kJ/m³ (400 Btu/ft³) and 18 600 kJ/m³ (500 Btu/ft³). Natural gas has a calorific value of 37 300 kJ/m³ (1 000 Btu/ft³).

In the case of electricity, the standard unit is the 'kilowatt hour' (kWh) which for practical purposes has a thermal value of 3 600 kJ.

Maximum Hourly Consumption

The maximum hourly fuel consumption is given by the expression:

For solid fuel

$$\text{Maximum hourly consumption in kg} = \frac{\text{Building heat loss in kW} \times 3\,600}{\text{Calorific value kJ/kg} \times \text{overall efficiency}}$$

For oil

$$\text{Maximum hourly consumption in litres} = \frac{\text{Building heat loss in kW} \times 3\,600}{\text{Calorific value kJ/litre} \times \text{overall efficiency}}$$

For gas

$$\text{Maximum hourly consumption in m}^3 = \frac{\text{Building heat loss in kW} \times 3\,600}{\text{Calorific value kJ/m}^3 \times \text{overall efficiency}}$$

For electricity

$$\text{Maximum hourly consumption in kW} = \frac{\text{Building heat loss in kW} \times 3\,600}{3\,600\text{ kJ} \times \text{overall efficiency}}$$

The Degree Day Method of Calculating Seasonal Fuel Consumption

Use of the degree day is the most satisfactory way of computing the probable fuel consumption over a given period.

The *degree day* can be defined as the difference between 60°F (5°F less than the usual room temperature in this country) and the daily mean outside temperature, when the daily mean outside temperature is below 60°F.

The temperature in an unheated building is usually 5°F higher than the outside temperature, and for this reason the *base temperature* of 60°F is used.

For the same period the weather can vary considerably year by year, and for this reason fuel consumption calculations should be based upon the average degree days over a number of years.

Table E.1 in Appendix E gives monthly average degree day figures for the period 1945–65 in the U.K., to a temperature base of 60°F, and Table E.2 for the period 1958–62 in Ireland.

Table 20 (parts A and B) gives monthly average degree days over the same periods to a temperature base of 15·6°C (60°F), derived from Tables E.1 and E.2 using the expression:

Degree day to base of 60°F × 0·56 = Degree day to base of 15·6°C.

The length of the heating season depends upon the building usage and location.

In most parts of England and Wales the approximate heating season for domestic, commercial, and public buildings is from 1 October to 30 April. Reference to columns 2 and 10 in Table 20 shows that such buildings in Scotland may well require heat during September and May.

In hospitals, old folk's hostels, nursing homes, infant welfare clinics, and buildings of this type throughout Britain, the heating season usually extends from 1 September to 31 May, and this is the only safe basis for estimating the probable fuel consumption of these buildings.

When using the Degree Day method it is assumed that the heating plant is fitted with effective thermostatic controls, with no possibility of sustained overheating of the building.

To determine the seasonal solid fuel consumption for a building in full use 24 hours per day, the appropriate Degree Day formula is:

$$\text{Tonnes of fuel for stated period} = \frac{\text{Degree days} \times 24}{16 \cdot 7} \times \frac{\text{kg/h}}{1\,000}$$

For oil-fired installations, the expression $\frac{\text{kg/hour}}{1\,000}$ becomes *litres per hour*; in the case of gas, *cubic metres per hour*; electricity, *units per hour*; the consumption being in these units.

Example. Consider a nursing home in Lancashire, the calculated hourly maximum fuel consumption for which is 51 kg per hour.

$$\text{Fuel used 1 Sept. to 31 May} = \frac{2\,260 \times 24}{16 \cdot 7} \times \frac{51\ \text{kg/h}}{1\,000} = 165 \text{ tonnes.}$$

Table 20. DEGREE DAYS
(A) Average for 20 year period 1945–65 in Great Britain.
Compiled from data issued by the Gas Council.
Base temperature = 15·6°C

Area	Sep.	Oct.	Nov.	Dec.	Jan.	Feb.	Mar.	Apl.	May	Total
Thames Valley	54	140	234	318	354	317	280	187	112	1 996
South Eastern	74	170	273	358	397	353	313	221	143	2 302
Southern	56	129	225	308	347	309	271	183	114	1 942
Western	47	113	200	269	309	286	255	181	116	1 776
Severn Valley	66	151	247	330	366	328	293	204	131	2 116
Midland	74	164	265	348	383	344	305	214	141	2 238
Lancashire	81	167	266	354	384	340	305	219	144	2 260
North Western	96	179	267	351	385	346	317	231	155	2 327
North Eastern	100	186	284	371	400	357	336	241	175	2 450
Yorkshire	78	169	270	350	388	343	317	213	140	2 268
Eastern	72	161	265	352	388	339	302	207	134	2 220
S.E. Scotland	100	183	277	354	388	353	331	241	179	2 406
West Scotland	106	197	298	378	413	363	334	239	162	2 490
East Scotland	104	192	295	375	405	353	331	237	167	2 459

(B) Average for 5 year period 1958–62 in Ireland.
Compiled from data issued by the Department of Transport and Power Meteorological
Service, Dublin. Base temperature = 15·6°C

Area	Sep.	Oct.	Nov.	Dec.	Jan.	Feb.	Mar.	Apl.	May	Total
Roches Point	54	125	213	275	288	241	258	197	135	1 786
Cahirciveen	50	118	198	254	264	218	230	177	121	1 630
Rosslare	48	122	210	281	302	253	273	204	141	1 834
Kilkenny	87	170	268	348	356	275	293	209	142	2 148
Shannon Airport	58	142	242	324	332	259	265	191	121	1 934
Birr	81	167	265	349	353	276	290	211	140	2 132
Dublin Airport	70	154	247	327	343	273	294	215	146	2 069
Mullingar	94	182	282	367	377	297	309	225	150	2 283
Claremorris	97	174	273	357	370	290	297	225	156	2 239
Clones	97	178	282	370	380	298	310	231	153	2 299
Belmullet	65	133	221	296	313	250	267	202	143	1 890
Malin Head	73	137	228	304	321	268	286	225	164	2 006

The consumption for the three coldest months, December, January, and February, will be:

$$\text{Fuel used Dec. to Feb. inclusive} = \frac{(354 + 384 + 340) \times 24}{16·7} \times \frac{51}{1\,000} = 78·7 \text{ tonnes.}$$

Hospitals and nursing homes require full heat over the 24-hour period. The majority of buildings, however, are heated intermittently, or at reduced temperatures during the night time between 6 p.m.

and 6 a.m. To allow for the resultant fuel economy the estimated consumption for full 24-hour heating, as calculated above, is multiplied by a suitable constant. For offices and most other buildings where the heating can be shut down at 6 p.m. until 6 a.m. and at weekend, the fuel saving is of the order of 45 per cent for oil or gas, and 40 per cent for solid fuel. The lower saving with solid fuel is due to banking the furnace when coke is used, and for periodic re-kindling during the shut-down period when using small coal with underfeed automatic stokers.

In those schools which are not in use for evening activities, and which normally close down at 4 p.m. until 6 a.m. and over the week-end, savings will be, oil or gas 50 per cent; for solid fuel 45 per cent is possible.

Where continuous heating at a reduced temperature during the unoccupied hours is practised (room temperature reduced from 18°C to 13°C), the savings are generally lower than for intermittent heating, and for the purpose of seasonal fuel prediction, savings with all fuels of 25 per cent in commercial buildings and up to 30 per cent in schools can be assumed.

Table 21 gives constants for intermittent and partial heating based upon the average savings.

Table 21. FUEL CONSUMPTION CONSTANTS (PERCENTAGES OF FULL HEATING CONSUMPTION) FOR INTERMITTENT AND CONTINUOUS REDUCED TEMPERATURE HEATING

	Commercial buildings		Schools	
	Intermittent heating Shut down 6 p.m. to 6 a.m.	*Continuous heating* Rooms at 13 C 6 p.m. to 6 a.m.	*Intermittent heating* Shut down 4 p.m. to 6 a.m.	*Continuous heating* Rooms at 13 C 4 p.m. to 6 a.m.
Coke	0·60	0·75	0·55	0·70
Coal	0·60	0·75	0·55	0·70
Oil	0·55	0·75	0·50	0·70
Gas	0·55	0·75	0·50	0·70

Example. From the data given in the foregoing pages, estimate the probable seasonal fuel consumption for an intermittently-heated office building in the Midlands, having an automatically coal-fired boiler, in charge of the office caretaker, and a calculated heat loss of 293 kW.

Calorific value of fuel, say 27 900 kJ/kg.
Overall efficiency = 60 per cent (from Table 19, Column 3).
Heating season, 1 October to 30 April.

Degree days Columns 3 to 9 inclusive, Table 20 = 2 023.

First, find the hourly consumption.

$$\text{Maximum hourly consumption} = \frac{\text{Building heat loss, kW} \times 3\,600}{\text{Calorific value, kJ/kg} \times \text{overall efficiency}}$$

$$= \frac{293 \text{ kW} \times 3\,600}{27\,900 \text{ kJ/kg} \times 0{\cdot}60}$$

$$= 63{\cdot}01 \text{ kg.}$$

$$\text{Seasonal fuel consumption in tonnes} = \frac{2\,023 \times 24}{16{\cdot}7} \times \frac{63{\cdot}01}{1\,000} \times 0{\cdot}60 = 109{\cdot}9 \text{ tonnes.}$$

In the above calculation no allowance has been made for the few holidays which occur at Christmas and Easter, when the offices will be closed. These only amount to 4 or 5 days at the most and in commercial buildings of this type the effect upon the fuel consumption estimate may be disregarded. With schools, however, which close for longer holiday periods, the degree days must be reduced if the heating system is shut down.

ESTIMATING THE ANNUAL FUEL CONSUMPTION OF HOT WATER SUPPLY INSTALLATIONS

The heat required for a hot water supply system consists of that for heating the water to the required temperature, usually 10°C to 66°C, and that due to pipe and other surface losses.

Unlike space heating, the hot water load is not a seasonal one, although in residential buildings, the consumption during the winter months tends to rise. Table 22 gives the average daily consumption per head of population in certain types of building, and can be used for estimating the annual hot water consumption.

The heat loss from the boiler, cylinder, and circulating mains in boiler room and trenches, wastage of hot water, the effects of poor maintenance and attention, and other utilisation losses, are allowed for in the overall efficiencies given in Table 19, p. 000.

Heat emission from circulating pipes in rooms must be added to that required for heating the water. The heat loss from a gravity secondary circulation will continue during banked-down periods when the building is not in use, unless of course the flow valve on the main circulation is operated each day, a duty which is liable to be neglected.

For intermittently-operated solid fuel systems (which are banked during the shut-down period), with gravity secondary circulations,

Table 22. AVERAGE DAILY CONSUMPTION OF HOT WATER

Type of building	Hot water at 66°C (150°F) consumed per day per resident; patient; resident staff; factory or office worker			
	LBs fitted with standard taps		LBs fitted with spray taps	
	litres	gallons	litres	gallons
Residential buildings, including boarding houses, children's homes, old peoples hostels, working girls hostels, working boys hostels, hotels, boarding schools, etc.	114	25	105	23
Commercial and public buildings, including offices, large shops and stores, police stations, fire stations	Women 19·5 Men 10·5	Women 4·3 Men 2·3	Women 10·5 Men 5·9	Women 2·3 Men 1·3
Ambulance stations	23	5	14	3
Factory and pit-head showers (spray type)	23	5		
Secondary schools	23	5	18	4
Primary schools	14	3	11	2·4
Hospitals:				
General	136	30		
Infectious	227	50		
Maternity	227	50		
Mental	91	20		
Nurses homes	136	30		
Blocks of flats:				
Low rental	68	15		
Medium rental	114	25		
High rental	136	30		

The consumptions for hospitals and blocks of flats are quoted from the *Guide to Current Practice* 1965 by permission of the Institution of Heating and Ventilating Engineers.

Note: The consumptions quoted include for ablutions, cooking, laundry and cleaning where appropriate.

it is therefore advisable to allow for the heat emission continuing during 24 hours per day. With oil, gas, and electric systems, after the initial cooling of the water left in the pipes, the heat loss will be reduced during the shut-down period. If the secondary circulation is pump-assisted, the heat loss can be limited to the hours of occupation by leaving the pump switched off during the shut-down period.

The daily fuel consumption for a hot water installation is given by:

$$\text{Daily fuel consumption} = \frac{\begin{matrix}\text{Daily heat to produce} \\ \text{hot water, kJ}\end{matrix} + \begin{matrix}\text{Daily heat emission from} \\ \text{circulating pipes in rooms, kJ}\end{matrix}}{\text{Calorific value of fuel, kJ/kg} \times \text{Overall efficiency}}$$

Example. Estimate the annual fuel consumption of a hot water supply system, installed in a children's home occupied by 50 children and 7 resident staff. The boiler is coke-fired by hand, and in the care of the gardener. Secondary system fitted with pump. In operation from 7 a.m. to 11 p.m. each day, 7 days per week, 48 weeks each year. (Children and staff at seaside for 4 weeks in summer.)

Hot water temperature = 66°C.
Cold water feed temperature = 10°C.
Hourly heat loss from pipes in rooms calculated at 2·931 kW.
Calorific value of coke = 27 900 kJ/kg.
Overall efficiency (Table 19, Column 3) = 50 per cent.

Daily hot water consumption per head = 114 litres at 66°C.
Daily hot water consumption in the home
   = 114 litres × 57
   = 6 498 litres.

Heat to raise 6 498 litres at 10°C to the storage temperature of 66°C, will be:

   6 498 litres × 1 kg/litre × (66°C—10°C) × 4·187
   = 6 498 × 56 × 4·187
   = 1 523 000 kJ/day.

The system operates for 16 hours per day, and the daily heat loss from secondary circulating pipes in the rooms, will be:

$$2\text{·}931 \text{ kW} \times 16 \times 3\,600 = 168\,900 \text{ kJ/day.}$$

$$\text{Daily fuel consumption} = \frac{1\,523\,000 \text{ kJ} + 168\,900 \text{ kJ}}{27\,900 \text{ kJ/kg} \times 0\text{·}50} = 121\text{·}3 \text{ kg.}$$

$$\text{Estimated annual fuel consumption} = \frac{121\text{·}3 \text{ kg} \times 7 \text{ days} \times 48 \text{ weeks}}{1\,000}$$

   = 40·71 tonnes   say 41 tonnes.

MAINTENANCE

The annual cost of maintaining the installation in good running order depends upon the type of plant, general design, annual running time, and the quality of day-to-day care and attention given to the equipment, especially in the boiler room. The effects of these variable influences are unpredictable, and it is usual to add a small percentage based upon the capital cost of the boiler room plant, to cover repairs, lubrication, and daily maintenance. The greater portion of maintenance expenditure is for repair work and running materials, such as lubricating oil, and minor replacements in the boiler room, or in the case of a ventilating system the plant room.

The annual cost of maintenance outside the boiler room will, to a large extent, depend upon the type of equipment installed in the building. Unit heaters and cabinet type air heaters require considerably more maintenance to ensure constant efficiency than free-standing radiators. Steam heating needs considerably more attention than hot water heating, and high temperature heating more than low temperature systems.

As already stated, a lot depends upon the daily care and attention given to the installation by the operator, and it is often noticed that although the annual maintenance costs of a certain plant are consistently high, due to frequent breakdowns, another exactly similar plant in the district, which was installed at the same time, remains trouble free. This leads one to the conclusion that although records

Table 23. APPROXIMATE ANNUAL MAINTENANCE COSTS FOR HEATING AND HOT WATER SUPPLY INSTALLATIONS, AS A PERCENTAGE OF CAPITAL COST OF BOILER HOUSE PLANT
(For estimating purposes only)

Type of heating surface in building	Percentage of plant capital cost to allow for annual maintenance
Steam-heated air heaters, or unit heaters	4
Water-heated air heaters, or unit heaters	3
Steam-heated convectors	3
Water-heated convectors	$2\frac{1}{2}$
Steam radiators	$2\frac{1}{2}$
Water radiators, low pressure	2
Water radiators, pressurised system	$2\frac{1}{2}$
Invisible ceiling heating (water)	$1\frac{1}{2}$
Floor heating (water	$1\frac{1}{2}$

of the maintenance costs of existing installations are a useful guide when estimating similar costs for new buildings and should be used whenever possible, such information is not wholly reliable, and should be used with care.

Where maintenance cost records are not available, Table 23 may be used to estimate the approximate annual maintenance cost.

Table 24 may be used for estimating the maintenance costs of installations employing independent gas and electric space and water heaters.

Table 24. APPROXIMATE ANNUAL MAINTENANCE COSTS FOR GAS AND ELECTRIC HEATING AND HOT WATER SUPPLY INSTALLATIONS USING INDEPENDENT HEATERS, AS A PERCENTAGE OF THE TOTAL CAPITAL COST OF THE INSTALLATION

(For estimating purposes only. Water heating by individual heaters)

Type of heating	Percentage of capital cost
Overhead radiant gas heating	$2\frac{1}{2}$
Gas convectors (flue type)	2
Gas convectors (flueless)	$1\frac{1}{2}$
Gas radiators	1
Gas warm-air units (flue type with fan-assisted air circulation)	3
Electric unit heaters	$1\frac{1}{2}$
Electric radiant, reflector type heaters	1
Electric convectors (natural air circulation)	1
Electric convectors (fan-assisted air circulation)	$1\frac{1}{2}$
Electric radiators (oil-filled)	$\frac{3}{4}$
Electric floor warming and ceiling heating	$\frac{1}{4}$
Electric block storage heating	1

VENTILATION SYSTEMS

The maintenance of a simple centrifugal fan-operated ventilation system is very low. Sheet metal ductwork and fittings will normally last a lifetime without skilled attention, and if the fan receives periodic inspection, cleaning, and lubrication, it, too, will run for many years without giving trouble. Some provision must be made, however, for renewal of bearings, starter maintenance, and the proper upkeep of the electrical wiring and control systems.

Air conditioning installations have a much higher maintenance factor, due to the nature of the equipment. Air filters, washers, heaters, humidifiers and refrigerating equipment need careful and constant maintenance.

Allow 2 per cent for simple mechanical ventilating systems and 4 per cent for air conditioning installations, based in each case upon the capital cost of the plant room equipment as installed.

ANNUAL COST OF ELECTRICITY USED IN PLANT ROOM

The cost of electricity for plant room lighting, pump motors, stoker motors and oil burners, and in the case of many oil-fired systems, electricity for maintaining the oil storage temperature, must be included in the estimate.

Lighting costs can be found by:

$$\frac{\frac{\text{Lamp watts}}{\text{installed}} \times \frac{\text{Estimated hours}}{\text{in use daily}} \times \frac{\text{Estimated number}}{\text{of days}}}{1\,000} \times \frac{\text{Cost}}{\text{per}}_{\text{unit}}$$

Annual cost of heating oil = Heater rating in kilowatts × Estimated hours in use daily × Estimated number of days × Cost per unit.

The electricity consumption of an electric motor is given by:

$$\text{kW consumed} = \frac{\text{kW power} \times 100}{\text{Motor efficiency}}$$

Table 25. AVERAGE FULL LOAD EFFICIENCIES OF A.C. ELECTRIC MOTORS

Motor power		Silent type		Industrial type	
kW	hp	1 phase Eff. %	3 phase Eff. %	1 phase Eff. %	3 phase Eff. %
0·187	¼	57	60	60	63
0·373	½	62	65	65	68
0·746	1	67	70	70	74
1·49	2	74	76	78	80
2·24	3	75	77	79	81
3·73	5	76	78	80	82
7·46	10	79	81	83	85
11·19	15	80	82	84	86
14·92	20	82	84	85	88

The approximate efficiencies at full load of single-phase and three-phase induction motors, at 50 hertz, and running at 1 500 rev/min, are given in Table 25.

The annual cost of electricity consumed by an electric motor will be:

$$\text{Annual cost} = \frac{\text{kW power} \times 100}{\text{Motor efficiency}} \times \frac{\text{Running time}}{\text{in hours}} \times \frac{\text{Unit cost of}}{\text{electricity}}$$

Example. Estimate the approximate annual cost of electricity for an industrial-type, 1·49 kW, three-phase motor.

Running time: 12 hours per day, 5 days per week for 30 weeks = 1 800 hours.
Motor efficiency = 80 per cent (Table 25).
Cost of electricity, per unit = 1p.

$$\text{Estimated annual cost} = \frac{1 \cdot 49 \times 100}{80} \times 1\ 800 \text{ hours} \times 1 \text{p}$$

$$= 1 \cdot 862 \times 1\ 800 \text{p}$$
$$= 3\ 353 \text{p} = £33 \cdot 53.$$

STANDING CHARGES

Labour for Attention to Plant

Standing charges have a considerable influence upon the annual cost, particularly in the case of the larger installations where a regular boiler or plant attendant, or attendants, are engaged.

The cost of labour for stoking, furnace cleaning, watching, cleaning and running adjustments to pumps, automatic stokers and oil burners, trimming fuel, removal of ashes and general care of plant, is a legitimate charge upon the installation.

In large buildings where a resident maintenance engineer is employed, a proportion of his time is charged to the heating and ventilation services.

Table 26. TIME ALLOWANCE FOR ATTENDANCE

Capacity of plant in operation		Daily hours for plant attendance				
kW	1 000 Btu/h	Coke hand-fired	Coal hopper-type auto-stoker	Coal bunker to boiler auto-stoker	Oil burner	Gas and electric boilers
147	500	2½	2	1½	1	½
293	1 000	3	2½	2	1½	½
440	1 500	3½	3	2½	2	¾
586	2 000	4½	3½	3	2	¾
879	3 000	5½	4½	4	2½	1
1 172	4 000	6½	5½	4	2½	1
1 465	5 000	8	6½	4	2½	1

For the great majority of the centrally-heated buildings, the work is undertaken by the caretaker and a charge should be made based upon the actual time spent on boiler room duties, which will vary according to the size and type of system. No hard and fast rule for estimating this labour can be given, and each installation must be assessed on its merits.

It should be stated, however, that the smallest of boilers whether hand-fired or automatic gas, oil, or solid fuel, needs some labour to keep it clean and working reasonably efficiently.

Table 26 is offered as an approximate guide when estimating the time for attendance to heating and hot water supply installations.

Repayment of Loans

Money for financing capital building and engineering products is usually borrowed. The rate of interest on the loan will be decided by the lender, who may be a bank, the Public Works Loan Board, or private finance houses.

The borrower will make provision to repay the capital with interest annually over a number of years.

Repayment of capital and interest on that part of the loan devoted to the purchase and installation of the engineering services and the boiler house becomes a standing charge upon the annual cost of these services.

The amount per £100 borrowed, to be paid annually to liquidate capital and interest, over the term of the loan, is given by:

$$P = \frac{R}{1 - \left(1 + \dfrac{R}{100}\right)^n}$$

Where P = The annual payment per £100 borrowed.
R = The percentage rate of interest.
n = The term of the loan in years.

Table 27 is based upon the above formula for an interest rate of 7 per cent.

An example will serve to show how this standing charge is determined.

Example. Suppose the estimated capital cost of the heating and hot water installation and boiler house of a building project, which is to be financed on a 20-year loan at 7 per cent, is £18 000. Calculate the standing charge.

Table 27

Term of loan in years	Amount of repayment per annum, per £100 borrowed
10	£14 23
15	10 98
20	9 43
30	8 06
40	7 50

From Table 27, the annual repayment sum per £100 of capital loan is £9·43.

The total annual sum payable on £18 000

$$= \frac{£18\,000}{100} \times £9\cdot43$$

$$= £1\,697\cdot00$$

This sum of £1 697·00 represents an annual standing charge on the installation, and as such must be included in the estimate of annual costs.

If a business organisation decides to finance a similar project from its own resources, an annual sum to cover depreciation of engineering plant and boiler house will be chargeable. In this case, each part of the project will be depreciated over a period determined by its expected useful life, as follows:

Boiler house   £3 600 at 40 years

Boiler room plant   £4 200 at 15 years

Radiators and installations in the building   £10 200 at 25 years

The annual sum to be allowed for depreciation will be:

Boiler house $\dfrac{£3\,600}{40\ \text{years}} = £90\cdot00$

Boiler room plant $\dfrac{£4\,200}{15\ \text{years}} = £280\cdot00$

Installations in the building $\dfrac{£10\,200}{25\ \text{years}} = £408\cdot00$

Total amount of standing charge for depreciation   £778·00

In this calculation no account has been taken of interest charges to compensate for the use of this capital had it been invested. In private enterprise such interest charges are invariably waived.

Insurance

Expenditure incurred by the owner for fire insurance of buildings and plant, and also workmen's compensation and employers' liability insurances, are charges on the upkeep of the building; where these apply to engineering services, including the labour for attendance and maintenance, they must be considered as an annual standing charge to be included in the estimate. The same applies to the employers' National Insurance contributions.

Insurance premiums for buildings and plant depend upon the amount insured and the nature of the risk, which in turn are governed by the value of buildings and plant and the use to which the buildings are put.

An insurance company will naturally ask a much bigger annual premium for a building in which inflammable goods are manufactured than for an office building of equal capital value.

When preparing his estimate of annual running costs the heating engineer should obtain the annual premium values from the owner's insurance company. Insurances covering steam boilers and other pressure vessels to satisfy the requirements of the Factory Acts, and any other breakdown insurance on sectional heating boilers, oil burners, automatic stokers, pumps, ventilating fans, etc., which the owner may decide to take up, should also be noted.

Many Local Authorities and commercial undertakings insure heating and ventilating plant against electrical and mechanical breakdowns, and the annual premium to cover this risk alone may, in a large boiler room, cost upwards of £100 per annum.

Establishment Charges

The engineering plant's contribution to the establishment charges of a large organisation is often neglected when considering running costs. Items are: telephone calls to manufacturers and contractors for spares and maintenance, orders for fuel, postage for engineering correspondence, the examination of invoices and accounts, payment of wages and accounts, and bank charges, and all the clerical work connected with modern business which is carried out by the clerical staff. That portion of the overhead expenses which contributes to the smooth operation of the engineering services should be included in the estimate.

Table 28. SUM TO BE INCLUDED TO COVER OFFICE ESTABLISHMENT CHARGES, WHEN ESTIMATING ANNUAL COSTS OF HEATING, HOT WATER SUPPLY, AND VENTILATION SERVICES

(For estimating purposes only)

Type of installation	Annual charge as a percentage of capital cost of boiler room plant as installed
Central scheme. Solid fuel or oil-fired	$\frac{1}{2}$
Central scheme gas boilers	$\frac{1}{4}$
Independent heaters for both space and water heating (no boiler plant)	Annual charge as a percentage of capital cost of installation
Gas heating	$\frac{1}{4}$
Electric heating	$\frac{1}{6}$
Electric floor and ceiling heating	$\frac{1}{8}$
Ventilating and air conditioning	Annual charge equal to $\frac{1}{4}$ per cent of capital cost of plant room equipment as installed

Percentage values for estimating this charge approximately are given in Table 28.

The Charge for Central Office Supervision

Where the building and services are under the permanent supervision of an architect or consulting engineer, as in the case of Government Departments, Local Authorities, and large business houses, the engineering installations are inspected at regular intervals by head office technical assistants.

Where such arrangements are known to be in operation, an approximate sum to cover the annual cost must be included. It is not possible to include a token sum for estimating purposes when dealing with a charge of this nature, due mainly to the varying distances from head office. The estimator will therefore take each case on its merits, and base his annual charge upon an agreed number of visits (usually four each year), taking into account travelling and subsistence expenses, and salary. Where the running and maintenance is directed from a central office, an appropriate annual charge to cover engineer's and drawing office costs will also be included in the estimate.

Appendix A

RECOMMENDED PUBLICATIONS

The specification writer and those concerned with estimating in the heating and ventilating industry will find the undermentioned publications useful. The prices are correct at the time of going to press.

Title	Available from	£	s	d	£	p
British Standards Yearbook	British Standards Institute, Sales Dept., Newton House, 101 Pentonville Road, London N.1	1	0	0	1	00
Daywork. HVCA/RICS Definition of Prime Cost	Heating and Ventilating Contractors Association, Coastal Chambers, 172 Buckingham Palace Rd., London S.W.1		1	6		7½
Heating and Ventilating Yearbook	Heating and Ventilating Contractors Association, Coastal Chambers, 172 Buckingham Palace Rd., London S.W.1	2	5	0	2	25
Heating and Ventilating Man's Safety Guide	Heating and Ventilating Contractors Association, Coastal Chambers, 172 Buckingham Palace Rd., London S.W.1		10	0		50
National Agreement (Labour)	Heating and Ventilating Contractors Association, Coastal Chambers, 172 Buckingham Palace Rd., London S.W.1		3	0		15
Guide to Estimating	Heating and Ventilating Contractors Association, Coastal Chambers, 172 Buckingham Palace Rd., London S.W.1	3	0	0	3	00
Identification of Ductwork. Code of Practice (DW/161)	H.V.C.A., Coastal Chambers, 172 Buckingham Palace Rd., London S.W.1		9	0		45

Title	Available from	Price				
		£	s	d	£	p
Standard range of rectangular ducts and fittings. DW/112, in Metric and Imperial units	H.V.C.A., Coastal Chambers, 172 Buckingham Palace Rd., London S.W.1		6	0		30
Specification for sheet metal ductwork (Metric) DW/121. Low velocity systems	H.V.C.A., Coastal Chambers, 172 Buckingham Palace Rd., London S.W.1	2	0	0	2	00
As above in Imperial units (DW/122B)	H.V.C.A., Coastal Chambers, 172 Buckingham Palace Rd., London S.W.1	2	0	0	2	00
As above. High velocity systems (DW/131). (Imperial units)	H.V.C.A., Coastal Chambers, 172 Buckingham Palace Rd., London S.W.1	1	0	0	1	00
A.I.H. Standard Tender Conditions For M/E Contracts	H.V.C.A., Coastal Chambers, 172 Buckingham Palace Rd., London S.W.1		2	6		$12\frac{1}{2}$
Min of PB and W. Standard Specification (M and E) No. 3. Heating, HWS, Steam and Gas	H.M.S.O., 49 High Holborn, London W.C.1		7	6		$37\frac{1}{2}$
Min of PB and W. Standard Specification (M and E) Electrical Installations. No. 1	H.M.S.O., 49 High Holborn, London W.C.1		7	6		$37\frac{1}{2}$
Industrial oil fuel delivery, storage and handling	On request from Shell-Mex and B.P. Ltd., Shell-Mex House, Strand, London W.C.2					
Storage and Handling of Domestic Oil Fuels	On request from Shell-Mex and B.P. Ltd., Shell-Mex House, Strand, London W.C.2					
Regulations for the Electrical Equipment of Buildings 14th Edition 1970 Metric units	Institution of Electrical Engineers, Savoy Place, London W.C.2	1	5	0	1	25

Title	Available from	Price				
		£	s	d	£	p
Information on Gas Installations	Gas Council, 59 Bryanston Street, London W1A 2AZ					
Information on Solid Fuel	Coal Utilisation Council, 19 Rochester Row, London S.W.1					
Information on Electricity	Electricity Council, 30 Millbank, London S.W.1					

The following list of publications deal with the change from the Imperial to the SI system of measurement.

Title	Available from	£	s	d	£	p
B.S. 350: Part 1: 1959 Conversion Factors and Tables	British Standards Institution, 101 Pentonville Road, London N.1		15	0		75
B.S. 350: Part 2: 1962 Conversion Factors and Tables	British Standards Institution, 101 Pentonville Road, London N.1	1	5	0	1	25
B.S. 350: Part 2: Supplement: 1967	British Standards Institution, 101 Pentonville Road, London N.1	1	0	0	1	00
B.S. Handbook No. 16. Metric Tables	British Standards Institution, 101 Pentonville Road, London N.1		3	6		$17\frac{1}{2}$
B.S. PD 5686: 1967. The Use of SI Units	British Standards Institution, 101 Pentonville Road, London N.1		2	0		10
B.S. 1957: 1953. Presentation of Numerical Values	British Standards Institution, 101 Pentonville Road, London N.1		5	0		25
B.S. 3763: 1964. International System (SI) Units	British Standards Institution, 101 Pentonville Road, London N.1		6	0		30

Title	Available from	£	s	d	£	p
B.S. 6031: 1968. The Use of the Metric System in the Construction Industry	British Standards Institution, 101 Pentonville Road, London N.1		7	6		37½
B.S. 1192: 1969. Recommendations for Building Drawing office Practice	British Standards Institution, 101 Pentonville Road, London N.1	1	4	0	1	20
B.S. 6030: 1967. Programme for the Change to the Metric System in the Construction Industry	British Standards Institution, 101 Pentonville Road, London N.1		5	0		25
I.H.V.E. Guide Book C: 1970	I.H.V.E., 49 Cadogan Square, London S.W.1	6	6	0	6	30
I.H.V.E. Guide Book A: 1971	I.H.V.E., 49 Cadogan Square London S.W.1	8	8	0	8	40
Change to Metric Manual	I.H.V.E., 49 Cadogan Square, London S.W.1		15	0		75

Addresses of Organisations from whom information useful to the estimator and designer, is obtainable.

Title	Address	Tel. No.
British Standards Institution	British Standards House, 2 Park Street, London W.1	01-629 9000
British Standards Institution Sales Dept.	Newton House, 101 Pentonville Road, London N.1	01-278 2161
Coal Utilisation Council	19 Rochester Row, London S.W.1	01-834 2339
Construction Industry Training Board	Radnor House, London Road, Norbury, London S.W.16	01-764 5060
British Coke Research Association	Wingerworth, Chesterfield, Derbyshire	0246 76821

Title	Address	Tel. No.
Copper Development Association	Orchard House Mutton Lane, Potters Bar, Herts.	Potters Bar 50815
Building Industry Research Station	Garston, Watford, Hertfordshire	Garston 47 74040
Heating and Ventilating Research Association	Old Bracknell Lane, Bracknell, Berkshire	0344 5071
Electricity Council	30 Millbank, London S.W.1	01-834 2333
Structural Insulation Association	32 Queen Ann Street, London W.1	01-580 7616
Gas Council	59 Bryanston Street, London W1A 2AZ	01-723 7030
Esso Petroleum Ltd.	Victoria Street, London S.W.1	01-834 6677
Heating and Ventilating Contractors Association	Coastal Chambers, 172 Buckingham Palace Road, London S.W.1	01-730 8245
Her Majesty's Stationery Office	Atlantic House, Holborn Viaduct, London E.C.1	01-248 9876
Shell-Mex and B.P. Ltd.	Shell-Mex House, Strand, London W.C.2	01-836 1234

Appendix B
IMPERIAL/SI CONVERSION FACTORS
SI/IMPERIAL CONVERSION FACTORS

Quantity	Imperial unit	SI unit		Conversion
Length	mile	kilometre (km)	1 mile 1 km	$=1{\cdot}609$ km $=6{\cdot}214\times10^{-1}$ miles
	ft	metre (m)	1 ft 1 m	$=3{\cdot}048\times10^{-1}$ m $=3{\cdot}281$ ft
	in	millimetre (mm)	1 in 1 mm	$=2{\cdot}54\times10^{1}$ mm $=3{\cdot}937\times10^{-2}$ in
Mass	ton	tonne	1 ton 1 tonne 1 tonne	$=1{\cdot}016$ tonnes $=1\,000$ kg $=9{\cdot}842\times10^{-1}$ ton
	cwt	tonne	1 cwt	$=5{\cdot}08\times10^{-2}$ tonne
	lb	kilogramme (kg)	1 lb 1 kg	$=4{\cdot}536\times10^{-1}$ kg $=2{\cdot}205$ lb
Time	h. min. s.	h. s.		
Area	ft²	m²	1 ft² 1 m²	$=9{\cdot}29\times10^{-2}$ m² $=1{\cdot}0764\times10^{1}$ ft²
	yd²	m²	1 yd² 1 m²	$=8{\cdot}361\times10^{-1}$ m² $=1{\cdot}196$ yd²
Volume	ft³	m³	1 ft³ 1 m³	$=2{\cdot}832\times10^{-2}$ m³ $=3{\cdot}5315\times10^{1}$ ft³
	yd³	m³	1 yd³ 1 m³	$=7{\cdot}646\times10^{-1}$ m³ $=1{\cdot}308$ yd³
	U.K. gallons	litres	1 gallon 1 litre	$=4{\cdot}546$ litres $=2{\cdot}20\times10^{-1}$ gallons
Temperature	°F	°C	°F °C	$=\frac{9}{5}°C+32$ $=(°F-32)\times\frac{5}{9}$
Mass flow rate	lb/h	kg/s	1 lb/h 1 kg/s	$=1{\cdot}260\times10^{-4}$ kg/s $=7\,937$ lb/h
Volume flow rate	gal/h	litre/s	1 gal/h 1 litre/s	$=1{\cdot}260\times10^{-3}$ litre/s $=7{\cdot}936\times10^{2}$ gal/h
	gal/min	litre/s	1 gal/min 1 litre/s	$=7{\cdot}577\times10^{-3}$ litre/s $=1{\cdot}32\times10^{2}$ gal/min

265

Quantity	Imperial unit	SI unit	Conversion	
Volume flow rate (contd.)	ft³/min	m³/s	1 ft³/min	$=4\cdot719\times10^{-4}$ m³/s
			1 m³/s	$=2\cdot119\times10^{3}$ ft³/min
	ft³/min	litre/s	1 ft³/min	$=4\cdot719\times10^{-1}$ litre/s
			1 litre/s	$=2\cdot119$ ft³/min
Pressure	lbf/in²	N/m²	1 lbf/in²	$=6\cdot895\times10^{3}$ N/m²
			1 N/m²	$=1\cdot450\times10^{-4}$ lbf/in²
			1 lbf/in²	$=6\cdot895$ kN/m²
			1 kN/m²	$=1\cdot450\times10^{-1}$ lbf/in²
	lbf/in²	bars	1 lbf/in²	$=6\cdot895\times10^{-2}$ bars
			1 bar	$=1\cdot450\times10^{1}$ lbf/in²
Operating or working pressure	ft H$_2$O	millibars (mbar)	1 ft H$_2$O	$=2\cdot989\times10^{1}$ mbar
			1 mbar	$=3\cdot346\times10^{-2}$ ft H$_2$O
	in H$_2$O	mbar	1 in H$_2$O	$=2\cdot491$ mbar
			1 mbar	$=4\cdot015\times10^{-1}$ in H$_2$O
Pressure drop per unit length	in H$_2$O/ft	N/m³ (N/m² m)	1 in H$_2$O/ft	$=8\cdot176\times10^{2}$ N/m³
			1 in H$_2$O/10 ft	$=8\cdot176\times10^{1}$ N/m³
			1 in H$_2$O/100 ft	$=8\cdot176$ N/m³
			1 N/m³	$=1\cdot223\times10^{-3}$ in H$_2$O/ft
			1 N/m³	$=1\cdot223\times10^{-2}$ in H$_2$O/10 ft
			1 N/m³	$=1\cdot223\times10^{-1}$ in H$_2$O/100 ft
Mass per unit length	lb/ft	kg/m	1 lb/ft	$=1\cdot488$ kg/m
			1 kg/m	$=6\cdot720\times10^{-1}$ lb/ft
Mass per unit area	lb/ft²	kg/m²	1 lb/ft²	$=4\cdot882$ kg/m²
			1 kg/m²	$=2\cdot048\times10^{-1}$ lb/ft²
Density (Specific mass)	lb/ft³	kg/m³ =g/litre	1 lb/ft³	$=1\cdot602\times10^{1}$ kg/m³ =g/litre
			1 kg/m³	$=1$ g/litre $=6\cdot241\times10^{-2}$ lb/ft³
	lb/gal	kg/m³ =g/litre	1 lb/gal	$=9\cdot978\times10^{1}$ kg/m³ =g/l
			1 kg/m³	$=1$ g/litre $=1\cdot002\times10^{-2}$ lb/gal
Specific volume	ft³/lb	m³/kg	1 ft³/lb	$=6\cdot243\times10^{-2}$ m³/kg
			1 m³/kg	$=1\cdot602\times10^{1}$ ft³/lb

Quantity	Imperial unit	SI unit		Conversion
Heat, quantity of	Btu	joule (J)	1 Btu 1 Btu 1 joule	$=1 \cdot 055 \times 10^3$ J $=1 \cdot 055$ kilojoule (kJ) $=9 \cdot 478 \times 10^{-4}$ Btu
Power, heat flow rate	Btu/h	Watts (W) joules per second	1 Btu/h 1 Btu/h 1 Watt 1 kW	$=2 \cdot 931 \times 10^{-1}$ W $=2 \cdot 931 \times 10^{-4}$ kW $=3 \cdot 412$ Btu/h $=3 \cdot 412 \times 10^3$ Btu/h
	horsepower hp	W	1 hp 1 hp 1 Watt 1 kW	$=7 \cdot 457 \times 10^2$ Watts (W) $=7 \cdot 457 \times 10^{-1}$ kW $=1 \cdot 341 \times 10^{-3}$ hp $=1 \cdot 341$ hp
Intensity of heat flow rate	Btu/h ft²	W/m²	1 Btu/h ft² 1 W/m²	$=3 \cdot 155$ W/m² $=3 \cdot 170 \times 10^{-1}$ Btu/h ft²
Specific energy (calorific value)	Btu/lb	J/kg	1 Btu/lb 1 Btu/lb 1 J/kg 1 kJ/kg	$=2 \cdot 326 \times 10^3$ J/kg $=2 \cdot 326$ kJ/kg $=4 \cdot 299 \times 10^{-4}$ Btu/lb $=4 \cdot 299 \times 10^{-1}$ Btu/lb
Volumetric calorific value	Btu/ft³	J/m³	1 Btu/ft³ 1 Btu/ft³ 1 Btu/ft³ 1 J/m³ 1 kJ/m³ 1 MJ/m³	$=3 \cdot 726 \times 10^4$ J/m³ $=3 \cdot 726 \times 10^1$ kJ/m³ $=3 \cdot 726 \times 10^{-2}$ MJ/m³ $=2 \cdot 684 \times 10^{-5}$ Btu/ft³ $=2 \cdot 684 \times 10^{-2}$ Btu/ft³ $=2 \cdot 684 \times 10^1$ Btu/ft³
	Btu/gal	kJ/litre	1 Btu/gal 1 kJ/litre	$=2 \cdot 320 \times 10^{-1}$ kJ/litre $=4 \cdot 310$ Btu/gal

CONVERSION FACTORS
Non-SI metric units to equivalent SI units

Quantity				
Pressure	mbar bar kgf/m² mm H₂O	N/m²	1 mbar 1 bar 1 kgf/m² 1 mm H₂O	$=100$ N/m² $=10^5$ N/m² $=9 \cdot 807$ N/m² $=9 \cdot 807$ N/m²
Heat	calorie (cal)	Joule (J)	1 cal	$=4 \cdot 187$ J
Heat flow rate	Calorie/ second (cal/s)	J/s =W	1 cal/s	$=4 \cdot 187$ W
	kilocalorie/ hour (kcal/h)		1 kcal/h	$=1 \cdot 163$ W

Quantity	Non SI Metric Unit	SI Unit	Conversion	
Specific energy (calorific value)	kcal/kg	kJ/kg	1 kcal/kg	$=4\cdot187$ kJ/kg
Calorific value (volume basis)	kcal/m³	kJ/m³	1 kcal/m³	$=4\cdot187$ kJ/m³
Intensity of heat flow rate	kcal/m² h	W/m²	1 kcal/m² h	$=1\cdot163$ W/m²

Appendix C

CONVERSION TABLES

DEGREES FAHRENHEIT TO DEGREES CELSIUS
(Figures in italics represent negative values on the Celsius scale)

Degrees F	0	1	2	3	4	5	6	7	8	9
	°C	°C	°C	°C	°C	°C	°C	°C	°C	°C
0	*17·8*	*17·2*	*16·7*	*16·1*	*15·6*	*15·0*	*14·4*	*13·9*	*13·3*	*12·8*
10	*12·2*	*11·7*	*11·1*	*10·6*	*10·0*	*9·4*	*8·9*	*8·3*	*7·8*	*7·2*
20	*6·7*	*6·1*	*5·6*	*5·0*	*4·4*	*3·9*	*3·3*	*2·8*	*2·2*	*1·7*
30	*1·1*	*0·6*	—	—	—	—	—	—	—	—
	0	1	2	3	4	5	6	7	8	9
30	—	—	0	0·6	1·1	1·7	2·2	2·8	3·3	3·9
40	4·4	5·0	5·6	6·1	6·7	7·2	7·8	8·3	8·9	9·4
50	10·0	10·6	11·1	11·7	12·2	12·8	13·3	13·9	14·4	15·0
60	15·6	16·1	16·7	17·2	17·8	18·3	18·9	19·4	20·0	20·6
70	21·1	21·7	22·2	22·8	23·3	23·9	24·4	25·0	25·6	26·1
80	26·7	27·2	27·8	28·3	28·9	29·4	30·0	30·6	31·1	31·7
90	32·2	32·8	33·3	33·9	34·4	35·0	35·6	36·1	36·7	37·2
100	37·8	38·3	38·9	39·4	40·0	40·6	41·1	42·7	42·2	42·8
110	43·3	43·9	44·4	45·0	45·6	46·1	46·7	47·2	47·8	48·3
120	48·9	49·4	50·0	50·6	51·1	51·7	52·2	52·8	53·3	53·9
130	54·4	55·0	55·6	56·1	56·7	57·2	57·8	58·3	58·9	59·4
140	60·0	60·6	61·1	61·7	62·2	62·8	63·3	63·9	64·4	65·0
150	65·6	66·1	66·7	67·2	67·8	68·3	68·9	69·4	70·0	70·6
160	71·1	71·7	72·2	72·8	73·3	73·9	74·4	75·0	75·6	76·1
170	76·7	77·2	77·8	78·3	78·9	79·4	80·0	80·6	81·1	81·7
180	82·2	82·8	83·3	83·9	84·4	85·0	85·6	86·1	86·7	87·2
190	87·8	88·3	88·9	89·4	90·0	90·6	91·1	91·7	92·2	92·8
200	93·3	93·9	94·4	95·0	95·6	96·1	96·7	97·2	97·8	98·3
210	98·9	99·4	100·0	100·6	101·1	101·7	102·2	102·8	103·3	103·9
220	104·4	105·0	105·6	106·1	106·7	107·2	107·8	108·3	108·9	109·4
230	110·0	110·6	111·1	111·7	112·2	112·8	113·3	113·9	114·4	115·0
240	115·6	116·1	116·7	117·2	117·8	118·3	118·9	119·4	120·0	120·6
250	121·1	—	—	—	—	—	—	—	—	—

$$F = (C \times 1·8) + 32$$

APPENDIX C
DEGREES CELSIUS TO DEGREES FAHRENHEIT

Degrees C	0	1	2	3	4	5	6	7	8	9
	°F	°F	°F	°F	°F	°F	°F	°F	°F	°F
0	32·0	33·8	35·6	37·4	39·2	41·0	42·8	44·6	46·4	48·2
10	50·0	51·8	53·6	55·4	57·2	59·0	60·8	62·6	64·4	66·2
20	68·0	69·8	71·6	73·4	75·2	77·0	78·8	80·6	82·4	84·2
30	86·0	87·8	89·6	91·4	93·2	95·0	96·8	98·6	110·4	102·2
40	104·0	105·8	107·6	109·4	111·2	113·0	114·8	116·6	118·4	120·2
50	122·0	123·8	125·6	127·4	129·2	131·0	132·8	134·6	136·4	138·2
60	140·0	141·8	143·6	145·4	147·2	149·0	150·8	152·6	154·4	156·2
70	158·0	159·8	161·6	163·4	165·2	167·0	168·8	170·6	172·4	174·2
80	176·0	177·8	179·6	181·4	183·2	185·0	186·8	188·6	190·4	192·2
90	194·0	195·8	197·6	199·4	201·2	203·0	204·8	206·6	208·4	210·2
100	212·0	213·8	215·6	217·4	219·2	221·0	222·8	224·6	226·4	228·2
110	230·0	231·8	233·6	235·4	237·2	239·0	240·8	242·6	244·4	246·2
120	248·0	249·8	—	—	—	—	—	—	—	—

$$C = (F - 32) \div 1·8$$

FRACTIONS OF AN INCH
With decimal and metric equivalents

Fraction			Decimal	Millimetres	Fraction			Decimal	Millimetres
		1/64	0·015625	0·397			33/64	0·515625	13·097
	1/32		0·03125	0·794		17/32		0·53125	13·494
		3/64	0·046875	1·191			35/64	0·546875	13·891
	1/16		0·0625	1·587		9/16		0·5625	14·287
		5/64	0·078125	1·984			37/64	0·578125	14·684
	3/32		0·09375	2·381		19/32		0·59375	15·081
		7/64	0·109375	2·778			39/64	0·609375	15·478
1/8			0·125	3·175	5/8			0·625	15·874
		9/64	0·140625	3·572			41/64	0·640625	16·272
	5/32		0·15625	3·969		21/32		0·65625	16·669
		11/64	0·171875	4·366			43/64	0·671875	17·066
	3/16		0·1875	4·762		11/16		0·6875	17·462
		13/64	0·203125	5·160			45/64	0·703125	17·859
	7/32		0·21875	5·556		23/32		0·71875	18·256
		15/64	0·234375	5·953			47/64	0·734375	18·653
1/4			0·25	6·349	3/4			0·75	19·049
		17/64	0·265625	6·747			49/64	0·765625	19·477
	9/32		0·28125	7·144		25/32		0·78125	19·844
		19/64	0·296875	7·541			51/64	0·796875	20·241
	5/16		0·3125	7·937		13/16		0·8125	20·637
		21/64	0·328125	8·333			53/64	0·828125	21·034
	11/32		0·34375	8·731		27/32		0·84375	21·431
		23/64	0·359375	9·128			55/64	0·859375	21·828
3/8			0·375	9·524	7/8			0·875	22·224
		25/64	0·390625	9·922			57/64	0·890625	22·622
	13/32		0·40625	10·319		29/32		0·90625	23·019
		27/64	0·421875	10·716			59/64	0·921875	23·416
	7/16		0·4375	11·112		15/16		0·9375	23·812
		29/64	0·453125	11·509			61/64	0·953125	24·209
	15/32		0·46875	11·906		31/32		0·96875	24·606
		31/64	0·484375	12·303			63/64	0·984375	25·003
1/2			0·50	12·699	1			1·00	25·400

FEET AND INCHES TO METRES

Feet	Inches											
	0	1	2	3	4	5	6	7	8	9	10	11
	m	m	m	m	m	m	m	m	m	m	m	m
0	—	0·0254	0·0508	0·0762	0·1016	0·1270	0·1524	0·1778	0·2032	0·2286	0·2540	0·2794
1	0·3048	0·3302	0·3556	0·3810	0·4064	0·4318	0·4572	0·4826	0·5080	0·5334	0·5588	0·5842
2	0·6096	0·6350	0·6604	0·6858	0·7112	0·7366	0·7620	0·7874	0·8128	0·8382	0·8636	0·8890
3	0·9144	0·9398	0·9652	0·9906	1·0160	1·0414	1·0668	1·0922	1·1176	1·1430	1·1684	1·1938
4	1·2192	1·2446	1·2700	1·2954	1·3208	1·3462	1·3716	1·3970	1·4224	1·4478	1·4732	1·4986
5	1·5240	1·5494	1·5748	1·6002	1·6256	1·6510	1·6764	1·7018	1·7272	1·7526	1·7780	1·8034
6	1·8288	1·8542	1·8796	1·9050	1·9304	1·9558	1·9812	2·0066	2·0320	2·0574	2·0828	2·1082
7	2·1336	2·1590	2·1844	2·2098	2·2352	2·2606	2·2860	2·3114	2·3368	2·3622	2·3876	2·4130
8	2·4384	2·4638	2·4892	2·5146	2·5400	2·5654	2·5908	2·6162	2·6416	2·6670	2·6924	2·7178
9	2·7432	2·7686	2·7940	2·8194	2·8448	2·8702	2·8956	2·9210	2·9464	2·9718	2·9972	3·0226
10	3·0480	3·0734	3·0988	3·1242	3·1496	3·1750	3·2004	3·2258	3·2512	3·2766	3·3020	3·3274
11	3·3528	3·3782	3·4036	3·4290	3·4544	3·4798	3·5052	3·5306	3·5560	3·5814	3·6068	3·6322
12	3·6576	3·6830	3·7084	3·7338	3·7592	3·7846	3·8100	3·8354	3·8608	3·8862	3·9116	3·9370
13	3·9624	3·9878	4·0132	4·0386	4·0640	4·0894	4·1148	4·1402	4·1656	4·1910	4·2164	4·2418
14	4·2672	4·2926	4·3180	4·3434	4·3688	4·3942	4·4196	4·4450	4·4704	4·4958	4·5212	4·5466
15	4·5720	4·5974	4·6228	4·6482	4·6736	4·6990	4·7244	4·7498	4·7752	4·8006	4·8260	4·8514
16	4·8768	4·9022	4·9276	4·9530	4·9784	5·0038	5·0292	5·0546	5·0800	5·1054	5·1308	5·1562
17	5·1816	5·2070	5·2324	5·2578	5·2832	5·3086	5·3340	5·3594	5·3848	5·4102	5·4356	5·4610
18	5·4864	5·5118	5·5372	5·5626	5·5880	5·6134	5·6388	5·6642	5·6896	5·7150	5·7404	5·7658
19	5·7912	5·8166	5·8420	5·8674	5·8928	5·9182	5·9436	5·9690	5·9944	6·0198	6·0452	6·0706
20	6·0960	6·1214	6·1468	6·1722	6·1976	6·2230	6·2484	6·2738	6·2992	6·3246	6·3500	6·3754

	0	1	2	3	4	5	6	7	8	9	10	11
30	9·1440	9·1694	9·1948	9·2202	9·2456	9·2710	9·2964	9·3218	9·3472	9·3726	9·3980	9·4243
40	12·1920	12·2174	12·2428	12·2682	12·2936	12·3190	12·3444	12·3698	12·3952	12·4206	12·4460	12·4714
50	15·2400	15·2654	15·2908	15·3162	15·3416	15·3670	15·3924	15·4178	15·4432	15·4686	15·4940	15·5194
60	18·2880	18·3134	18·3388	18·3642	18·3896	18·4150	18·4404	18·4658	18·4912	18·5166	18·5420	18·5674
70	21·3360	21·3614	21·3868	21·4122	21·4376	21·4630	21·4884	21·5138	21·5392	21·5646	21·5900	21·6154
80	24·3840	24·4094	24·4348	24·4602	24·4856	24·5110	24·5364	24·5618	24·5872	24·6126	24·6380	24·6624
90	27·4320	27·4574	27·4828	27·5082	27·5336	27·5590	27·5844	27·6098	27·6352	27·6606	27·6860	27·7114
100	30·4800											

METRES TO FEET

Metres	0	1	2	3	4	5	6	7	8	9
0	—	3·281	6·562	9·843	13·123	16·404	19·685	22·966	26·247	29·528
10	32·808	36·089	39·370	42·651	45·932	49·213	52·493	55·774	59·055	62·336
20	65·617	68·898	72·179	74·459	78·740	82·021	85·302	88·583	91·864	95·144
30	98·425	101·706	104·987	108·268	111·549	114·829	118·110	121·391	124·672	127·953
40	131·234	134·515	137·795	141·076	144·357	147·638	150·919	154·200	157·480	160·761
50	164·042	167·323	170·604	173·885	177·166	180·446	183·727	187·008	190·289	193·570
60	196·851	200·131	203·412	206·693	209·974	213·255	216·536	219·816	223·097	226·378
70	229·659	232·940	236·221	239·502	242·782	246·063	249·344	252·625	255·906	259·187
80	262·467	265·748	269·029	272·310	275·591	278·872	282·152	285·433	288·714	291·995
90	295·276	298·557	301·838	305·118	308·399	311·680	314·961	318·242	321·523	324·803
100	328·08									

SQUARE FEET TO SQUARE METRES

ft^2	0	1	2	3	4	5	6	7	8	9
					Square metres					
0	—	0·09290	0·18581	0·27871	0·37161	0·46452	0·55742	0·65032	0·74322	0·83613
10	0·92903	1·02193	1·11484	1·20774	1·30064	1·39355	1·48645	1·57935	1·67225	1·76516
20	1·85806	1·95096	2·04387	2·13677	2·22967	2·32258	2·41548	2·50838	2·60129	2·69419
30	2·78709	2·87999	2·97290	3·06580	3·15870	3·25161	3·34451	3·43741	3·53032	3·62322
40	3·71612	3·80902	3·90193	3·99483	4·08773	4·18064	4·27354	4·36644	4·45935	4·55225
50	4·64515	4·73806	4·83096	4·92386	5·01676	5·10967	5·20257	5·29547	5·38838	5·48128
60	5·57418	5·66709	5·75999	5·85289	5·94579	6·03870	6·13160	6·22450	6·31741	6·41031
70	6·50321	6·59612	6·68902	6·78192	6·87482	6·96773	7·06063	7·15353	7·24644	7·33934
80	7·43224	7·52515	7·61805	7·71095	7·80386	7·89676	7·98966	8·08256	8·17547	8·26837
90	8·36127	8·45418	8·54708	8·63998	8·73289	8·82579	8·91869	9·01159	9·10450	9·19740
100	9·29030	9·38321	9·47611	9·56901	9·66192	9·75482	9·84772	9·94063	10·0335	10·1264

SQUARE METRES TO SQUARE FEET

m^2	0	1	2	3	4	5	6	7	8	9
					Square feet					
0	—	10·764	21·528	32·292	43·056	53·820	64·583	75·347	86·111	96·875
10	107·639	118·403	129·167	139·931	150·695	161·459	172·223	182·986	193·750	204·514
20	215·278	226·012	236·806	247·570	258·334	269·098	279·862	290·626	301·389	312·153
30	322·917	333·681	344·445	355·209	365·973	376·737	387·501	398·265	409·029	419·793
40	430·556	441·320	452·084	462·848	473·612	484·376	495·140	505·904	516·668	527·432
50	538·196	548·959	559·723	570·487	581·251	592·015	602·779	613·543	624·307	635·071
60	645·835	656·599	667·362	678·126	688·890	699·654	710·418	721·182	731·946	742·710
70	753·474	764·238	775·002	785·765	796·529	807·293	818·057	828·821	839·585	850·349
80	861·113	871·877	882·641	893·405	904·168	914·932	925·696	936·460	947·224	957·988
90	968·752	979·516	990·280	1001·04	1011·81	1022·57	1033·34	1044·10	1054·86	1065·63
100	1076·39	1087·15	1097·92	1108·68	1119·45	1130·21	1140·97	1151·74	1162·50	1173·27

CUBIC FEET TO CUBIC METRES

ft³	0	1	2	3	4	5	6	7	8	9
	m³	m³	m³	m³	m³	m³	m³	m³	m³	m³
0	—	0·0283	0·0566	0·0850	0·1133	0·1416	0·1699	0·1982	0·2265	0·2549
10	0·2832	0·3115	0·3398	0·3681	0·3964	0·4248	0·4531	0·4814	0·5097	0·5380
20	0·5663	0·5947	0·6230	0·6583	0·6796	0·7079	0·7362	0·7646	0·7929	0·8212
30	0·8495	0·8778	0·9061	0·9345	0·9628	0·9911	1·0194	1·0477	1·0760	1·1044
40	1·1327	1·1610	1·1893	1·2176	1·2459	1·2743	1·3026	1·3369	1·3592	1·3875
50	1·4158	1·4442	1·4725	1·5008	1·5291	1·5574	1·5857	1·6141	1·6424	1·6707
60	1·6990	1·7273	1·7556	1·7840	1·8123	1·8406	1·8689	1·8972	1·9255	1·9539
70	1·9822	2·0105	2·0388	2·0671	2·0954	2·1238	2·1521	2·1804	2·2087	2·2370
80	2·2653	2·2937	2·3220	2·3503	2·3786	2·4069	2·4352	2·4636	2·4919	2·5202
90	2·5485	2·5768	2·6051	2·6335	2·6618	2·6901	2·7184	2·7467	2·7750	2·8034
100	2·8317	—	—	—	—	—	—	—	—	—

CUBIC METRES TO CUBIC FEET

m³	0	1	2	3	4	5	6	7	8	9
	ft³	ft³	ft³	ft³	ft³	ft³	ft³	ft³	ft³	ft³
0	—	35·3148	70·6295	105·9443	141·2590	176·5738	211·8885	247·2033	282·5181	317·8328
10	353·1476	388·4623	423·7771	459·0918	494·4066	529·7214	565·0361	600·3509	635·6656	670·9804
20	706·2951	741·6099	776·9247	812·2394	847·5542	882·8689	918·1837	953·4984	988·8132	1024·1280
30	1059·4427	1094·7575	1130·0722	1165·3870	1200·7017	1236·0165	1271·3313	1306·6460	1341·9608	1377·2755
40	1412·5903	1447·9050	1483·2198	1518·5346	1553·8493	1589·1641	1624·4788	1659·7936	1695·1083	1730·4231
50	1765·7379	1801·0526	1836·3674	1871·6821	1906·9969	1942·3116	1977·6264	2012·9411	2048·2559	2083·5707
60	2118·8854	2154·2002	2189·5149	2224·8297	2260·1444	2295·4592	2330·7740	2366·0887	2401·4035	2436·7182
70	2472·0330	2507·3477	2542·6625	2577·9773	2613·2920	2648·6068	2683·9215	2719·2363	2754·5510	2789·8658
80	2825·1806	2860·4953	2895·8101	2931·1248	2966·4396	3001·7543	3037·0691	3072·3839	3107·6986	3143·0134
90	3178·3281	3213·6429	3248·9576	3284·2724	3319·5872	3354·9019	3390·2167	3425·5314	3460·8462	3496·1609
100	3531·47	—	—	—	—	—	—	—	—	—

GALLONS TO LITRES

gal	0	1	2	3	4	5	6	7	8	9
	l	*l*	*l*	*l*	*l*	*l*	*l*	*l*	*l*	*l*
0	—	4·546	9·092	13·638	18·184	22·730	27·276	31·822	36·368	40·914
10	45·460	50·006	54·552	59·098	63·643	68·189	72·735	77·281	81·827	86·373
20	90·919	95·465	100·011	104·557	109·103	113·649	118·195	122·741	127·287	131·833
30	136·379	140·925	145·471	150·017	154·563	159·109	163·655	168·201	172·747	177·293
40	181·839	186·384	190·930	195·476	200·022	204·568	209·114	213·660	218·206	222·752
50	227·298	231·844	236·390	240·936	245·482	250·028	254·574	259·120	263·666	268·212
60	272·758	277·304	281·850	286·396	290·942	295·488	300·034	304·580	309·125	313·671
70	318·217	322·763	327·309	331·855	336·401	340·947	345·493	350·039	354·585	359·131
80	363·677	368·223	372·769	377·315	381·861	386·407	390·953	395·499	400·045	404·591
90	409·137	413·683	418·229	422·775	427·321	431·866	436·412	440·958	445·504	450·050
100	454·596	—	—	—	—	—	—	—	—	—

LITRES TO GALLONS

litres	0	1	2	3	4	5	6	7	8	9
	gal	*gal*	*gal*	*gal*	*gal*	*gal*	*gal*	*gal*	*gal*	*gal*
0	—	0·2200	0·4400	0·6600	0·8800	1·1000	1·3199	1·5398	1·7598	1·9798
10	2·1998	2·4197	2·6397	2·8597	3·0797	3·2996	3·5196	3·7396	3·9596	4·1795
20	4·3995	4·6195	4·8395	5·0594	5·2794	5·4994	5·7194	5·9393	6·1593	6·3793
30	6·5993	6·8192	7·0392	7·2592	7·4792	7·6991	7·9191	8·1391	8·3591	8·5790
40	8·7990	9·0190	9·2390	9·4589	9·6789	9·8989	10·1189	10·3388	10·5588	10·7788
50	10·9988	11·2187	11·4387	11·6587	11·8787	12·0986	12·3186	12·5386	12·7586	12·9785
60	13·1985	13·4185	13·6385	13·8584	14·0784	14·2984	14·5184	14·7384	14·9583	15·1783
70	15·3983	15·6183	15·8382	16·0582	16·2782	16·4982	16·7181	16·9381	17·1581	17·3781
80	17·5980	17·8180	18·0380	18·2580	18·4779	18·6979	18·9179	19·1379	19·3578	19·5778
90	19·7978	20·0178	20·2377	20·4577	20·6777	20·8977	21·1176	21·3376	21·5576	21·7776
100	21·9975	—	—	—	—	—	—	—	—	—

POUNDS TO KILOGRAMMES

lb	0	1	2	3	4	5	6	7	8	9
	kg	*kg*	*kg*	*kg*	*kg*	*kg*	*kg*	*kg*	*kg*	*kg*
0	—	0·4535	0·9071	1·3607	1·8143	2·2679	2·7215	3·1751	3·6287	4·0823
10	4·5359	4·9895	5·4431	5·8967	6·3503	6·8039	7·2575	7·7111	8·1647	8·6183
20	9·0718	9·5254	9·9790	10·4326	10·8862	11·3398	11·7934	12·2470	12·7006	13·1542
30	13·6078	14·0614	14·5150	14·9686	15·4221	15·8757	16·3293	16·7829	17·2365	17·6901
40	18·1437	18·5973	19·0509	19·5045	19·9581	20·4117	20·8653	21·3188	21·7724	22·2260
50	22·6796	23·1332	23·5868	24·0404	24·4940	24·9476	25·4012	25·8548	26·3084	26·7620
60	27·2155	27·6691	28·1227	28·5763	29·0299	29·4835	29·9371	30·3907	30·8443	31·2979
70	31·7515	32·2051	32·6587	33·1122	33·5658	34·019	34·4730	34·9266	35·3802	35·8338
80	36·2874	36·7410	37·1946	37·6482	38·1018	38·5554	39·0089	39·4625	39·9161	40·3697
90	40·8233	41·2769	41·7305	42·1841	42·6377	43·0913	43·5449	43·9985	44·4521	44·9057
100	45·3592	—	—	—	—	—	—	—	—	—

KILOGRAMMES TO POUNDS

kg	0	1	2	3	4	5	6	7	8	9
	lb	*lb*	*lb*	*lb*	*lb*	*lb*	*lb*	*lb*	*lb*	*lb*
0	—	2·204	4·409	6·613	8·818	11·023	13·227	15·432	17·637	19·841
10	22·0462	24·250	26·455	28·660	30·864	33·069	35·273	37·478	39·683	41·887
20	44·0924	46·297	48·502	50·706	52·911	55·116	57·320	59·525	61·729	63·934
30	66·139	68·343	70·548	72·753	74·957	77·162	79·366	81·571	83·776	85·980
40	88·185	90·389	92·594	94·799	97·003	99·208	101·413	103·617	105·822	108·026
50	110·231	112·436	114·640	116·845	119·050	121·254	123·459	125·663	127·868	130·073
60	132·277	134·482	136·686	138·891	141·096	143·300	145·505	147·710	149·914	152·119
70	154·324	156·528	158·733	160·937	163·142	165·347	167·551	169·756	171·960	174·165
80	176·370	178·574	180·779	182·984	185·188	187·393	189·597	191·802	194·007	196·211
90	198·416	200·620	202·825	205·030	207·234	209·439	211·644	213·848	216·053	218·258
100	220·462	—	—	—	—	—	—	—	—	—

PRESSURE (STRESS)

POUNDS FORCE PER SQUARE IN (lbf/in²) TO KILONEWTONS PER SQUARE METRE (kN/m²)

	10	20	30	40	50
lbf/in² to kN/m²	68·948	137·90	206·84	275·79	344·74

	60	70	80	90	100
lbf/in² to kN/m²	413·69	482·63	551·58	620·53	689·48

Auxiliary Table

	1	2	3	4	5	6	7	8	9
lbf/in² to kN/m²	6·89476	13·7895	20·6843	27·5790	34·4738	41·3685	48·2633	55·1581	62·0528

KILONEWTONS PER SQUARE METRE (kN/m²) TO POUNDS FORCE PER SQUARE INCH (lbf/in²)

	10	20	30	40	50
kN/m² to lbf/in²	1·4504	2·9008	4·3511	5·8015	7·2519

	60	70	80	90	100
kN/m² to lbf/in²	8·7023	10·153	11·603	13·053	14·504

Auxiliary Table

	1	2	3	4	5	6	7	8	9
kN/m² to lbf/in²	0·14504	0·29008	0·43511	0·58015	0·72519	0·87023	1·01526	1·16030	1·30534

BRITISH THERMAL UNITS PER HOUR (Btu/h) TO KILOWATTS (kW)

1000 Btu/h	0	1	2	3	4	5	6	7	8	9
					Kilowatts					
0	—	0·2931	0·5862	0·8793	1·1724	1·4655	1·7586	2·0517	2·3448	2·6379
10	2·931	3·224	3·517	3·810	4·103	4·397	4·690	4·983	5·276	5·569
20	5·862	6·155	6·741	7·034	7·328	7·621	7·914	8·207	8·500	8·793
30	8·793	9·086	9·379	9·672	9·965	10·259	10·552	10·845	11·138	11·431
40	11·724	12·017	12·310	12·603	12·896	13·190	13·483	13·776	14·069	14·362
50	14·655	14·948	15·241	15·534	15·827	16·121	16·414	16·707	17·000	17·293
60	17·586	17·879	18·172	18·465	18·758	19·052	19·445	19·738	20·031	20·324
70	20·517	20·810	21·103	21·396	21·689	21·983	22·276	22·569	22·862	23·155
80	23·448	23·741	24·034	24·327	24·620	24·914	25·207	25·500	25·793	26·086
90	26·379	26·672	26·965	27·258	27·551	27·845	28·138	28·431	28·724	29·017
100	29·310	29·603	29·896	30·189	30·482	30·776	31·069	31·362	31·675	31·968

KILOWATTS (kW) TO BRITISH THERMAL UNITS PER HOUR (Btu/h)

kW	0	1	2	3	4	5	6	7	8	9
					1000 Btu/h					
0	—	3·412	6·824	10·236	13·648	17·060	20·472	23·884	27·296	30·708
10	34·12	37·532	40·944	44·356	47·768	51·180	54·592	58·004	61·416	64·828
20	68·24	71·652	75·064	78·476	81·888	85·300	88·712	92·124	95·536	98·948
30	102·36	105·77	109·18	112·60	116·01	119·42	122·83	126·24	129·66	133·07
40	136·48	139·89	143·30	146·72	150·13	153·54	156·96	160·37	163·79	167·20
50	170·60	174·01	177·42	180·84	184·25	187·66	191·07	194·48	197·89	201·31
60	204·72	208·13	211·54	214·96	218·37	221·78	225·19	228·60	232·02	235·43
70	238·84	242·25	245·66	249·08	252·49	255·90	259·31	262·72	266·14	269·55
80	272·96	276·37	279·78	283·20	286·61	290·02	293·43	296·84	300·26	303·67
90	307·08	310·49	313·90	317·32	320·73	324·14	327·55	330·96	334·78	337·79
100	341·20	344·61	348·02	351·44	354·85	358·26	361·67	365·08	368·50	371·91

HORSEPOWER TO KILOWATTS

Kilowatts

hp	0	1	2	3	4	5	6	7	8	9
0	—	0·7457	1·4914	2·2371	2·9828	3·7285	4·4742	5·2199	5·9656	6·7113
10	7·4570	8·2027	8·9484	9·6941	10·440	11·186	11·931	12·677	13·423	14·168
20	14·914	15·660	16·405	17·151	17·897	18·643	19·388	20·134	20·880	21·625
30	22·371	23·117	23·862	24·608	25·354	26·100	26·845	27·591	28·337	29·082
40	29·828	30·574	31·319	32·065	32·811	33·557	34·302	35·048	35·794	36·539
50	37·285	38·031	38·776	39·522	40·268	41·014	41·759	42·505	43·251	43·996

KILOWATTS TO HORSEPOWER

Horsepower

kW	0	1	2	3	4	5	6	7	8	9
0	—	1·3410	2·6820	4·0231	5·3641	6·7051	8·0461	9·3872	10·728	12·069
10	13·410	14·751	16·092	17·433	18·774	20·115	21·456	22·797	24·138	25·479
20	26·820	28·162	29·503	30·844	32·185	33·526	34·867	36·208	37·549	38·890
30	40·231	41·572	42·913	44·254	45·595	46·936	48·277	49·618	50·959	52·300

Appendix D

NOTES ON DECIMAL CURRENCY

In Great Britain, £.*s.d.* currency has been replaced by a decimal system which employs two units, the 'pound' for which the £ symbol is retained, and the 'new penny', abbreviated 'p'. The new system is generally referred to as the £p system.

The pound is divided into 100 pence instead of 240 pence in the £.*s.d.* system, and the lowest valued coin in circulation is the half-penny ($\frac{1}{2}$p).

This new system came into operation on 15 February 1971, and where monetary data and information, e.g. records of past expenditure, quotations, estimates, tenders, payments, purchases, and agreements relating to business conducted prior to that date, have to be referred to, conversion from £.*s.d.* to £p becomes necessary. One new penny is equal to $240/100 = 2\cdot 4d$ and therefore the basis of conversion will be as follows:

£.*s.d.*   £p
1 pound (£1. 0. 0.) = 1 pound (£1·00)

1 shilling (1*s*) $= \dfrac{12d}{2\cdot 4} = 5$p

1 penny (1*d*) $= \dfrac{1d}{2\cdot 4} = 0\cdot 4167$p

Example 1

Convert £750. 14s. 10d. to £p.

Solution

```
£750.  0s.  0d.              = £750·00
14s. × 5 = 70p              =    0·70
10d. × 0·4167 = 4·167p  =    0·04
£750.  14s.  10d.           = £750·74
```

Example 2

The 1970 catalogue trade price for a certain type and size of pipe bracket is quoted at 10d. each. Allowing for an increase in price of 10 per cent, estimate the probable cost (in £p), of 100 brackets.

Solution

10d. × 0·4167	= 4·167p
Add 10 per cent	= 0·4167p
Current price	= 4·5837p

Cost of 100 brackets = 4·5837 × 100 = 458·37p
= £4·5837
= £4·58½

Example 3

The weekly contribution for adult males to cover National Insurance, and Redundancy Fund is 94·5p. Find the hourly amount to include in the labour charge to cover this item, for one pair of men, based on a 40 hour week.

Solution

Weekly rate per pair = 94·5p × 2 = £1·89.

Hourly charge in estimate = $\dfrac{189p}{40}$ = 4·725p.

Example 4

The time for a pair of men to complete a proposed installation is estimated at 218 hours. Calculate the net amount to include in the tender for National Insurance, and Redundancy Fund contributions.

Sum to be included 4·725p × 218 = 1030p
= £10·30.

For internal accounting, job estimating, estimating running costs, and engineering costing generally, the £p system has pronounced advantages.

In most internal accounting processes it will aid calculation to express the ½p decimally as £0·005, rather than as a vulgar fraction, and as the above examples show, decimal fractions of lower value than the ½p (0·50p) must be used.

When pricing goods for retail sale, and in all transactions with the general public, the new halfpenny is expressed as a vulgar fraction, viz. ½p.

Booklets giving full information on all aspects of decimal currency, and coinage prepared by the Decimal Currency Board and the Central Office of Information, may be purchased from Her Majesty's Stationery Office, 49 High Holborn, London W.C.1.

Appendix E

IMPERIAL DEGREE DAY TABLES

Table E.1. DEGREE DAYS
(Average for 20-year period 1945-65 in Great Britain)
(Compiled from data issued by the Gas Council)
(Base temperature, 60°F)

Area	Sept.	Oct.	Nov.	Dec.	Jan.	Feb.	Mar.	Apl.	May	Total
No.										
1 Thames Valley	97	252	424	572	636	569	504	335	201	3 590
2 South Eastern	133	306	490	644	713	635	561	397	254	4 133
3 Southern	100	232	404	553	622	552	486	326	204	3 479
4 Western	84	201	357	480	552	511	455	323	207	3 170
5 Severn Valley	118	270	441	588	653	585	524	364	234	3 777
6 Midland	132	292	473	622	684	615	554	383	251	4 006
7 Lancashire	145	299	475	633	686	607	545	391	257	4 038
8 North Western	172	319	476	626	687	618	566	413	277	4 154
9 North Eastern	179	333	507	662	714	638	600	431	313	4 377
10 Yorkshire	140	302	482	642	693	612	566	380	250	4 067
11 Eastern	128	288	473	629	692	605	539	369	240	3 963
12 S.E. Scotland	178	327	495	633	693	630	591	430	319	4 296
13 West Scotland	190	351	532	675	737	648	597	427	289	4 446
14 East Scotland	185	342	526	670	724	630	591	424	299	4 391

Table E.2. DEGREE DAYS
(Average for 5-year period 1958–62 in Ireland)
(Compiled from data issued by the Department of Transport
and Power Meteorological Service, Dublin)
(Base temperature, 60°F)

15 Roches Point	96	223	381	494	514	430	460	352	241	3 191
16 Cahirciveen	89	211	353	453	471	389	410	316	216	2 908
17 Rosslare	85	218	376	501	539	452	487	365	252	3 275
18 Kilkenny	156	304	479	622	636	491	523	374	253	3 838
19 Shannon Airport	104	253	433	568	592	462	473	341	216	3 442
20 Birr	144	299	473	617	631	493	517	377	250	3 801
21 Dublin Airport	125	275	441	584	613	487	525	384	260	3 694
22 Mullingar	168	324	503	656	674	531	552	402	268	4 078
23 Claremorris	173	310	488	638	661	517	531	401	278	3 997
24 Clones	172	318	503	660	678	532	554	412	274	4 103
25 Belmullet	116	238	395	528	559	447	477	360	255	3 375
26 Malin Head	130	244	408	541	574	479	511	402	293	3 582

Appendix F

SCHEDULE OF RATES

NOTES ON THE COMPILATION OF SCHEDULE RATES FOR MEASURED WORK FOR THE PRICING OF CONTRACT VARIATIONS

Where the contract provides for a Schedule of Rates for the pricing of variations, this will take the form given in Chapter 2.

In the original tender, the labour item includes wages, statutory on-costs, costs to comply with the National Agreement between Union and Employers, and where applicable the costs of training (C.I.T.B. levy), and apprentice day release.

A percentage is then added to these estimated labour costs to cover the contractor's profit and overheads. The percentage to add, will, as explained in Chapter 1, depend upon trading conditions at the time of tendering, and also upon the firm's financial and trading position.

For the purpose of the examples in Chapter 5 of this book, a margin of 15 per cent was used to cover normal overheads and profit.

When preparing a Schedule of Rates to be used in assessing the cost of variations, the margin included in the tender (in this case, 15 per cent on labour and materials) must be increased to cover the Main Contractor's Discount, and cartage and handling of the extra materials which may be needed, and also an additional margin to cover the extra supervisory work on site and in the office, consequent upon the variation to the scheme.

Taking the Church Hall Scheme as an example, the labour charge is calculated as follows (see page 171):

Wages	£196·00
Statutory and National Agreement Commitments	£52·11
Sub-contract conditions	£37·22
Wear and tear of tools	£5·88
	£291·21
say	£291

and the net labour cost will be:

$$\text{Net labour cost} = \frac{£291}{196h} = £1·485 \simeq £1·50 \text{ per pair/hour.}$$

As stated, the normal margin, in this case 15 per cent on labour and materials, must be increased to cover the main contractor's discount, cartage and handling of materials, and supervision resulting from the variation order. In the tender summary (see page 000) the first two of these items amount to:

Cartage of materials	£16·80
Main contractor's discount	£26·59
	£43·39
say	£43

and the net cost of materials and wages (page 171) is:

Materials	£538·41
Wages	£196·00
	£734·41
say	£734

Margin to add to cover cartage of materials, and extra supervision and office work,

$$=\frac{£43 \times 100}{£734}=5·859 \simeq 6 \text{ per cent.}$$

For extra supervision and office work add 4 per cent, which for the purpose of compiling the Schedule of Rates for this particular job, gives a margin to be added to materials and labour costs of:

$$15+6+4=25 \text{ per cent}$$

and the labour charge for Schedule compilation, will be:

Net labour cost item=£1.50 per pair/hour
Add 25 per cent   =£0·375
                     £1·875

say £2·00 per pair/hour

Two examples follow, to show in tabular form, the calculation of schedule rates based upon a labour charge of £2·00 per pair/hour, and net cost of materials.

Supply and fix on the ground floor of a new building, black medium weight steel tubing to B.S. 1387, with screwed and socketed joints, and brackets (Brackets fixed by builder).

Size		Labour		Materials		
(1)		(2)	(3)	(4)	(5)	(6)
		Time fixing (Chapter 4 Section 1)	Labour fixing cost at £2·00 per pair hour	Net cost per metre run	Add 25% to Col. 4 (Rounded off)	Schedule rate per metre run (Col. 3 + Col. 5)
mm	in	hours	£	£	£	£
150	6	1·00	2·00	2·10	2·62	4·62
100	4	0·60	1·20	1·05	1·31	2·51
80	3	0·42	0·84	0·85	1·06	1·90
65	2½	0·35	0·70	0·66	0·83	1·53
50	2	0·30	0·60	0·43	0·54	1·14
40	1½	0·24	0·48	0·34	0·43	0·91
32	1¼	0·20	0·40	0·30	0·38	0·78
25	1	0·19	0·38	0·25	0·31	0·69
20	¾	0·18	0·36	0·14	0·18	0·54
15	½	0·15	0·30	0·12	0·15	0·45

Supply and fit at low level on the ground floor of a new building, bends as specified.

Size		Labour		Materials		
(1)		(2)	(3)	(4)	(5)	(6)
		Time fixing (Chapter 4 Section 1)	Fixing cost at £2·00 per hour pair	Net cost per bend	Add 25% to Col. 4 (Rounded off)	Schedule rate per bend (Col. 3 + Col. 5)
mm	in	hours	£	£	£	£
150	6	2·50	5·00	13·50	16·88	21·88
100	4	1·50	3·00	2·50	3·13	6·13
80	3	1·00	2·00	1·20	1·50	3·50
65	2½	0·90	1·80	0·96	1·20	3·00
50	2	0·80	1·60	0·50	0·63	2·23
40	1½	0·65	1·30	0·29	0·36	1·66
32	1¼	0·55	1·10	0·25	0·31	1·41
25	1	0·50	1·00	0·16	0·20	1·20
20	¾	0·40	0·80	0·13	0·16	0·96
15	½	0·35	0·70	0·10	0·13	0·83

The reader will note that the labour item in the above tabular calculations refers to work at low level on the ground floor of a new building.

For work under other conditions, the time rate should be adjusted as indicated in the table footnotes, Chapter 4, e.g. for fixing steel pipes and fittings on the third floor of a new building, the fixing time factor, Column 2 above, is increased by 15 per cent.

INDEX